COMMUNITY GEOGRAPHY

GIS in Action

KIM ZANELLI ENGLISH • LAURA S. FEASTER

ESRI Press

REDLANDS, CALIFORNIA

ESRI
   Community Geography: GIS in Action
   ISBN 1-58948-023-6

First printing June 2003.

Printed in the United States of America.

Library of Congress Cataloging-in-Publication Data
English, Kim.
      Community geography : GIS in action / Kim Zanelli English, Laura S. Feaster.
         p.   cm.
      Summary: Provides real case studies, hands-on exercises, and practical tips
   for using geographical information systems to learn about and make a
   difference in one's own community.
      Includes bibliographical references.
      ISBN 1-58948-023-6 (pbk. : alk. paper)
      1. Geographic information systems—Juvenile literature. [1. Geographic
   information systems. 2. Geography.] I. Feaster, Laura S. II. Title.
   G70.212.E54 2003
   910'.285—dc21   2003009456

Published by ESRI, 380 New York Street, Redlands, California 92373-8100.

Books from ESRI Press are available to resellers worldwide through Independent Publishers Group (IPG). For information on volume discounts, or to place an order, call IPG at 1-800-888-4741 in the United States, or at 312-337-0747 outside the United States.

Acknowledgments

Many people have had a hand in the development of this book. Counting all the students and teachers involved in the seven case studies, more than two hundred individuals played a part! Our biggest thank you goes out to all those students, teachers, and community partners who had the heart to use GIS to better their community. We are delighted to showcase the remarkable accomplishments of these young people and their teachers, and to draw attention to their creativity, dedication, and enthusiasm for GIS and community action. These students clearly show it is not necessary to be a GIS professional to embark upon a GIS project that provides great benefit to the community.

From start to finish, Lyn Malone, Anita M. Palmer, and Christine L. Voigt made key contributions to this book, for which we are deeply grateful. They were involved in the planning stages, they provided exercises and "on your own" ideas for each of the modules, and they reviewed the entire book. Their frequent feedback and insights about GIS education were essential. We thank them for their dedication to GIS and to the success of this project.

Heartfelt thanks are due to Claudia Naber, our editor with ESRI Press, for her role in making this book the best it could be. Her patience, wisdom, and light spirit got us over the inevitable rough spots with ease and gave the project that extra polish. Thank you to Angela Lee of the ESRI Schools and Libraries program who reviewed the entire book and provided technical assistance. Her attention to detail, familiarity with the intended audience of the book, and willingness to answer peculiar questions at a moment's notice were critical and much appreciated.

For their review of the entire book prior to publication, we'd like to thank Dr. Joseph Kerski, geographer for the U.S. Geological Survey; Gerry Bell, GIS teacher at Port Colborne High School in Port Colborne, Ontario, Canada; and Dr. Al Lewandowski of Port Huron Middle School in Port Huron, Michigan. In addition, thanks to Makram Murad-al-shaikh and Tim Ormsby at ESRI for their helpful comments.

Obtaining data and creating the data CD would not have been possible if it weren't for the technical assistance of a few key people. Thanks to Erin Kelly of ESRI–Canada, and to Scott McNair, Angela Lee, and Rick Schneblin of ESRI, for their expertise. We also thank the following people at ESRI for their assistance: George Dailey and Charlie Fitzpatrick with the ESRI Schools and Libraries program, and Gary Amdahl, Richard Greene, David Boyles, and Steve Hegle of ESRI Press.

This book would not have become a reality if it weren't for the support of ESRI. Thank you to Nick Frunzi, Judy Boyd, and Christian Harder of ESRI Educational Services for giving us the creative freedom we needed to produce our best work.

Finally, we'd like to thank Jack Dangermond for suggesting a book such as this one. The story of "Jack's trees" reminded us that a community project, even a small one, can have major, lasting significance in a young person's life.

Foreword

If I could do anything, I would want to inspire young people. And a part of me believes I can do anything, and I want young people to believe that about themselves too. This book is for them, and for active citizens. There is a great world out there, and if you get involved, you can shape it to reflect your own ideals and passions. GIS is the tool that can help you understand your environment—and with understanding, the sky's the limit.

I was lucky—I received excellent educational opportunities throughout my life for which I am grateful because I believe they inspired me to envision creating a company like ESRI. But before that happened, I was getting involved in projects like the ones described in this book, and that was before GIS was around to help. When I was nineteen, I saw an article in the local paper for a "Redlands Beautification Contest" the night before the submissions were due. At the time, I was in college studying to be a landscape architect so I stayed up all night with my brother, and our ideas ended up winning. Today the trees that we planted on State Street have formed a wonderful canopy that still shades shoppers in the Redlands town center. That was one of the first times I ever made a difference in my community, and I liked it. I learned something simple but profound from that experience—I understood clearly that the world is shaped by seemingly small ideas that are acted upon. It sounds too simple, but it's the truth—we have a good idea and we act on it—that is how we create a better world.

This book is filled with stories like that—about students and their teachers teaming with local businesses to learn about their environment in nontraditional ways. The students asked questions, they acquired resources to study their environment, they explored geographic information, and they analyzed geographic data. Finally, they took the knowledge they had gained and acted in their communities on behalf of their communities.

There are endless geographic questions to ask—you have them in your imaginations and you only need the chance to ask them, to explore and analyze them, and to act on them in order to reach others and to make a difference in your world.

I hope you'll go for it.

Jack Dangermond

How to use this book

This book contains stories and tools to inspire you to make a difference in your community using geographic information systems (GIS). Various sections of this book are written to guide you through the process of completing your own community GIS project. Before you begin completing the exercises or planning your own GIS project, make sure you have the following:

- A computer (Apple® Macintosh® or PC running Microsoft® Windows®)
- ESRI® ArcView® 3.x software
- An introductory knowledge of GIS or ArcView

HOW THE BOOK IS ORGANIZED

The book's design allows you to skim the book and jump right to the module that interests you the most. Each individual module is a complete unit, so you can skip around or read the book in chronological order. You may want to start with the warm-up exercise in module 1, "GIS basics." This exercise uses community data such as streets, landmarks, and business locations to create a visitor map.

Modules 2 through 8 are based on real-life community projects done by students across North America. Each of these modules has three parts: a case study, a GIS exercise, and an "on your own" section. You can quickly determine which part of a module you're in by consulting the header at the top of each page:

"On your own: Project planning" is an expanded "on your own" section, with information that applies to all GIS projects, independent of their topic.

Case study

Case study

The case studies showcase the GIS work done by youth in their communities. Each case study describes how students partnered with local organizations to use GIS for action. They are illustrated with photographs and maps from the actual projects. Each case study describes the challenges students faced in their project and the solutions they developed, allowing you to see through example how GIS can apply to a variety of community issues. The geographic inquiry steps taken by the students are summarized in a table at the end of each case study, allowing a quick reference to the main steps of each project.

Exercise

Exercise

Each step-by-step exercise is an adaptation of the case study and is based on the case-study data. They are written for educational purposes only. In essence, each exercise is a community GIS miniproject that follows the geographic inquiry process and helps you to better understand how to use geographic information systems. For example, in the module 6 exercise, "Use buffers to identify eligible school-bus riders," you start with a table of student addresses and street data, and conclude by making a recommendation to the school principal on which students are eligible to ride the bus to school and which are not. Each geographic inquiry step is noted in the margins.

Exercises include detailed instructions on GIS analysis, eight unique data sets, useful illustrations, and questions designed to help develop broad-based problem-solving skills.

Each question is noted by a pencil icon (✏).

Answers to all exercise questions are found in *Community Geography: GIS in Action Teacher's Guide.*

The exercise data and projects are located on the CD–ROM that comes with this book. All of the exercises written in this book have been tested on and function within Windows and Macintosh environments. Because Windows is the recommended platform for all ESRI® software products, all graphics and instructions in the exercises refer to the Windows platform. If you are using a Macintosh computer, refer to the Macintosh technical guide located at the back of the book for a description of differences in the operation of ArcView on the Windows and Macintosh platforms.

On your own

On your own

The "on your own" sections build on the information you learned in the case study and in the exercise. They provide a road map to completing a similar project in your own community. For example, in module 2 ("Reducing crime"), the "on your own" section provides information about working with police departments and analyzing crime data, while in module 7 ("Protecting the community forest"), the "on your own" section contains information about working with a local arborist and the public works department to collect and analyze data on trees.

Although each module's case study is unique, they are all organized around the geographic inquiry process, a five-step method that provides an ideal framework around which to organize a GIS project. This section is organized around the five-step geographic inquiry process. Detailed within each step of the geographic inquiry process are tips for how to complete a project in your community.

GEOGRAPHIC INQUIRY STEP	TIPS INCLUDED
Ask a geographic question	Contains suggestions on how to formulate geographic questions centered on the case-study topic.
Acquire geographic resources	Provides information on the types of data necessary for the GIS project and lists possible places to find it.
Explore geographic data	Contains technical information regarding the types of data used in the GIS project and suggestions on specific data-exploration steps.
Analyze geographic information	Includes tips on different analysis techniques and steps.
Act on geographic knowledge	Provides practical ideas on how to act on your analysis results.

On your own: Project planning

Many components of GIS project management apply to all GIS projects, regardless of their topic. This section explores those components such as data-use restrictions and permission, metadata, and creating an excellent public presentation of your project results. Based on the five-step geographic inquiry process, this section of the book follows the same structure of the "on your own" sections of each module. It is a must-read for anyone seeking to do a GIS project in his or her own community.

Resources

As you work through the modules of the book, be sure to use the additional resources included at the end. The back of the book contains data-installation guides for Windows and Macintosh users. It also includes a Macintosh technical guide. The data license agreement explains the permitted uses of the data on the CD–ROM. The resources and references section contains print and media resources for each module.

In addition, the book has its own Web site *(www.esri.com/communitygeography)*, complete with Internet resources, a discussion forum, and other valuable information related to this book.

Teacher's guide

If you are an educator, you will be interested in the teacher's guide that complements this book. It contains a lesson plan and an "on your own" project guide for each module. All of the lessons and "on your own" projects are correlated to National Geography Standards and National Science and Technology Standards. All answers to exercise questions are located in an answer key at the back of the teacher's guide. More information about *Community Geography: GIS in Action Teacher's Guide* can be found on the Community Geography Web site.

Now that you're familiar with this book's design, find a topic that interests you or that is important in your community. Start there to use GIS to make a difference.

Contents

Maps have been used throughout history to help us explore, inform, and find our way. In this module you will use GIS to create an informative city map. The purpose of the exercise is to help you review some basic ArcView skills that you'll be expected to know for the other community-based GIS exercises in the book.

EXERCISE

Explore and label community features data for a city visitors map

In this exercise, you will review how to use GIS to label features, explore attribute data, and observe spatial relationships in the city map. You will create and print a layout that will be used by visitors to the city.

Explore and label community features data for a city visitors map

There are many types of maps you can use to get information. Topographic maps, for example, show the elevation of a place, and political maps show boundaries between countries. When making a map, you need to carefully consider what type of map serves your purposes and which elements should be shown.

To complete this exercise you will use basic GIS skills to create a map, making it most useful to a specific audience. You will add data to the map, find and identify information on the map, and determine appropriate symbology. Finally, you will create a presentation layout of the finished map and print it.

The ⇨ icon indicates questions to be answered. Write your answers on a separate sheet of paper.

ASK

A group of American history teachers from New England is planning an educational trip to the presidential libraries and memorials in California. One stop on their trip is the Lincoln Memorial Shrine at the A. K. Smiley Library in Redlands. It is the only museum, archive, and library dedicated to Abraham Lincoln west of Springfield, Illinois. You are a member of a Redlands community group that has volunteered to use its GIS expertise to prepare informational maps for the teachers' visit to this shrine.

1 Double-click the ArcView icon on your computer's desktop to start ArcView. If you don't have an icon on your desktop, open ArcView from the Start menu.

2 From the File menu, click Open Project. (If the Welcome to ArcView GIS window displays, choose Open an existing project.)

3 Navigate to the exercise data folder *(C:\esri\comgeo\module1)* and open *map.apr.* Resize your View window so the map is larger.

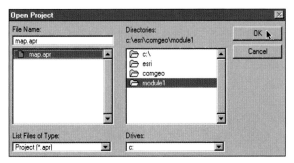

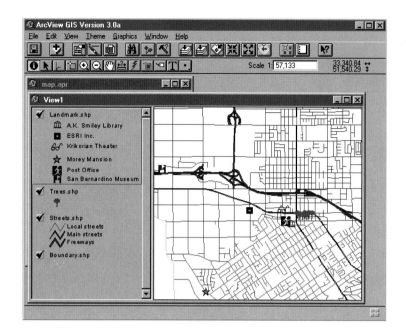

A map of streets and landmarks in Redlands, California, displays. The table of contents for the map symbols and data is presented at the left of the map. The data layers are called *themes* and the four themes in this map are Landmark.shp, Trees.shp, Streets.shp, and Boundary.shp.

3a What colors are the streets on your map?

3b What business is represented by the eyeglasses symbol?

3c What landmark is next to the A. K. Smiley Library?

Now that you have familiarized yourself with the map, you are ready to dig deeper. One of the powerful features of GIS is the ability to look at the information that is connected to each feature in the map. Each theme or layer of information in the table of contents has a separate table of data attached to that theme (called an attribute table).

4 Click the Identify tool and position the cursor over the symbol that represents ESRI, Inc., in the map. Click the symbol.

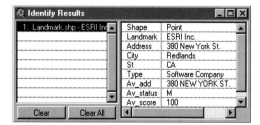

An Identify Results window appears. This window displays the name of the feature you clicked in the left column (ESRI, Inc.). Information about that feature is in the right column.

 4a What type of landmark is ESRI, Inc.?

ESRI is the company that develops and distributes the GIS software that is used in all the ArcView projects in this book.

5 **Click the symbol for the A. K. Smiley Library on the map.**

Information about the A. K. Smiley Library appears highlighted in the Identify Results window.

 5a What type of landmark is the A. K. Smiley Library?

 5b What is the address of the A. K. Smiley Library?

 5c Why is this location on the map important to know?

6 **Close the Identify Results window.**

In addition to historical locations and points of interest, visitors to Redlands will need to find lodging and places to eat. You will add those additional layers of data to the map you're preparing for them.

ACQUIRE

7 **Click the Add Theme button. Navigate to the exercise data folder (C:\esri\comgeo\module1). Click once on coffee_dessert.shp in the list of files at the left to highlight it. Hold down the Shift key and click once on lodging.shp. With both files highlighted, click OK.**

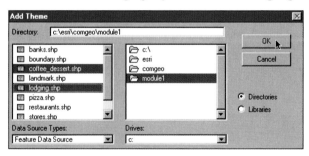

EXPLORE

8 Click the small box to the left of Lodging.shp. A check mark appears in the box, indicating that Lodging.shp is "turned on" and displayed in the view.

※ **NOTE:** The symbol may be a different color on your computer than in the picture below.

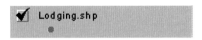

8a What happened to the map when you turned on Lodging.shp?

8b Does the Coffee_dessert.shp theme appear on your map?

8c Why or why not?

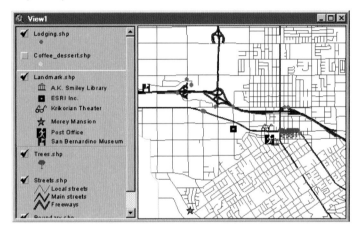

As the map designer, you have an obligation to make your map as readable and usable as possible for its intended audience. When Lodging.shp was added, ArcView assigned a small, randomly colored dot to represent lodging locations. You will change that dot to a more appropriate symbol.

9 In the table of contents, double-click the Lodging.shp theme. In the Legend Editor symbol box, double-click the colored dot symbol.

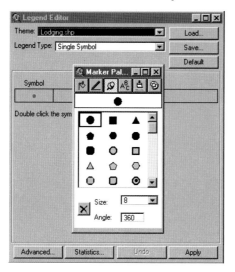

10 Scroll down the list of symbols until you find the roadside sign symbols. Click the symbol that depicts a bed. Leave the Size as 8 and Angle as 360.

✴ **NOTE:** The symbol might look better at size 10 or 14 depending on your version of ArcView.

11 If necessary, click the Color Palette button and change the color to blue. Click Apply in the Legend Editor. Close the Legend Editor and the Marker Palette.

✐ 11a What change did you see on the map?

12 Turn on Coffee_dessert.shp. Use the procedure from steps 9 and 10 to change the symbol for Coffee_dessert.shp to a **coffee cup**, the size to **12**, and the color to **dark red**.

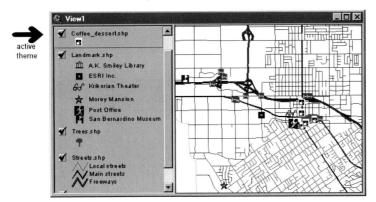

active theme

Notice that when you clicked the Coffee_dessert.shp theme, it changed to look like a raised button. This tells you the theme is active. When you want to perform analysis on a specific theme, you first make that theme active.

Now you will change the names of the themes in your legend to make them more readable.

13 Click the Theme Properties button. Change the theme name to **Coffee and Dessert Shops**. Click OK.

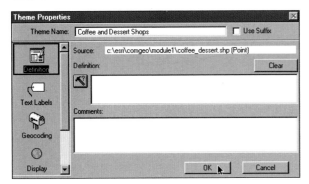

13a Where does the new title Coffee and Dessert Shops appear on your screen?

13b Why is it sometimes helpful to change the theme name from the file name to something else?

14 Using the process of making a theme active (single-click it) and clicking the Theme Properties button, change the other theme names to match the graphic below.

The visitors are interested in staying in hotels and motels in Redlands. Some are interested in staying close to the downtown area of Redlands where the Lincoln Memorial Shrine is located, and others have a preference for a particular hotel chain (Best Western).

You will explore the list of hotels and motels and map them. First, you will look at the attribute table, which contains all the data for that theme.

15 Make the Hotels and Motels theme active. Click the Open Theme Table button.

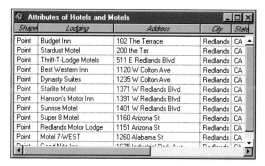

16 Look for a Best Western, then click it. Move the attribute table so you can see the entire map.

16a What happened to the map?

16b What Redlands landmark is close to the Best Western?

When you click a record in the attribute table, that record is selected in the table and its corresponding feature is selected in the view.

17 Make the View window active and click the Select Feature tool. Click the symbol for the hotel nearest the Lincoln Memorial Shrine. Look at the attribute table.

17a What is the name and address of the hotel you selected?

18 Close the Attribute Table. Click the Clear Selected Features button.

19 Click the Zoom In tool. Position your cursor at the top left of the downtown area of Redlands as shown in the picture below. Click and drag the mouse down and to the right. When your box is approximately the same size and shape as in the picture, release the mouse button.

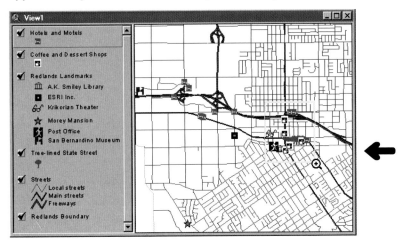

✳ **CAUTION:** When you zoom, make sure to keep the mouse very still as you click or you may accidentally drag a tiny box and the map will zoom in too much. If this happens, go to the toolbar and click the Zoom to Previous Extent button once.

To create an informative map for the Redlands visitors, you will label the different features.

20 From the Window menu, click Show Symbol Window and click the Font button. Change the Size to **9**. Close the window.

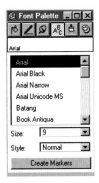

The font size is changed for all labeling you will do until you change the Symbol Window font size again.

21 From the Theme menu, click Auto-label. In the Auto-label window, click the box to Allow Overlapping Labels and click OK.

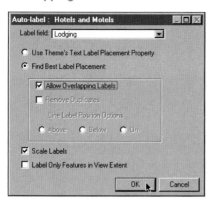

The labels appear in the view. Some overlap. You will move these labels so you can read them.

22 Click the Pointer tool and then click Thrift-T-Lodge Motels. Black "handles" around the label indicate that the label is selected. Move the label so that it's to the right of the symbol. See the picture below.

23 Click the Pan tool. Click the map and move it to the left so you can read the entire Thrift-T-Lodge Motels label and still see the symbol for the Krikorian Theater.

24 Click the Pointer tool and move the labels for the Budget Inn and the Stardust Motel in line with their symbols.

25 Use the same labeling process you learned in steps 20 and 21 to label Coffee and Dessert Shops.

✷ **NOTE:** If you do not allow overlapping labels, it's possible that some features may not be labeled at all.

Depending on the size of your view window, you may have noticed that some of the labels appear green. ArcView automatically assigns any overlapping labels a green color. Once the label is in a nonoverlapping position, it's easy to change the color.

26 Move any overlapping labels so they can be clearly read and are not overlapping. Hold the Shift key down and click each of the green labels so the black handles appear around them. From the Window menu, click Show Symbol Window. Click the Color Palette button. Select Text from the Color drop-down menu and click black. All of the selected labels change to black in the view.

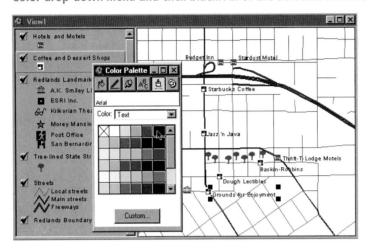

27 Close the Color Palette. Click once in the view and the black handles will disappear.

In addition to labeling hotels and motels and coffee and dessert shops, you will need to label major streets in downtown Redlands. To label some of the city streets, you will use the Label tool.

28 Make the Streets theme active. Click the Label tool and click the freeway southeast of the Stardust Motel in the map.

The label I10 appears in the same spot as where you clicked the freeway. Use the Pointer tool to move the label into a position where it can be read easily.

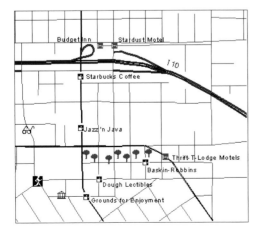

29 Use the labeling procedure you learned in step 28 to label the following streets:

Colton Redlands Orange State Eureka Citrus Vine

When you're finished with the labeling, your map should look similar to the one below:

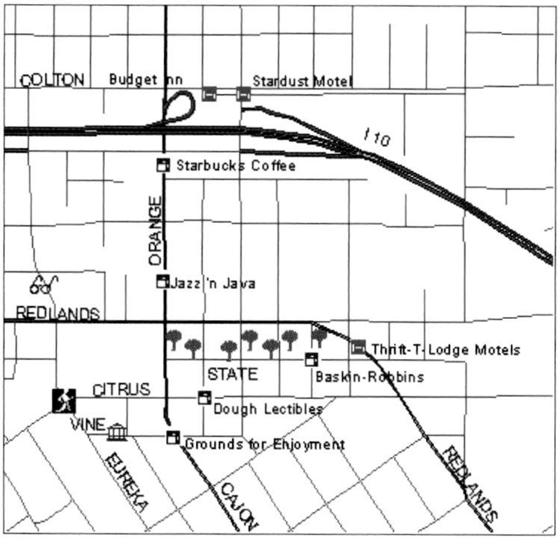

As an added tool for the Redlands visitors, you will create a 0.5-mile circle around the Lincoln Memorial Shrine. This will show which locations are within walking distance of the Lincoln Memorial Shrine.

ANALYZE

30 From the Draw tools drop-down list, click the Draw Circle tool. Click the symbol for the A. K. Smiley Library. Drag your cursor out from the symbol until you've created a circle with a 0.5-mile radius.

NOTE: The radius measurement is located on the bottom left of the View window. If necessary, zoom out and reposition your map so the entire circle is visible and centered in the view.

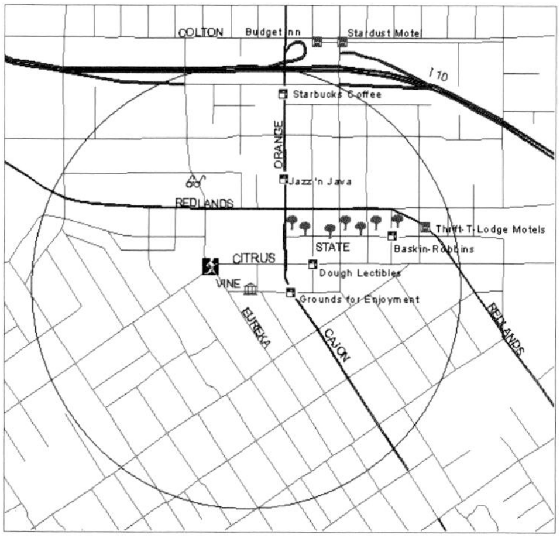

31 Click the project window title bar. From the File menu, choose Save Project As. If you are in a classroom environment, ask your instructor for directions on how to rename your project and where to save it (e.g., **map_abc.apr** where "abc" represents your initials). Save your project.

Now that your map is clearly symbolized and labeled, you are ready to prepare it for printing. In the next steps, you will create a layout to display your completed map. Your layout will include the map, a title, a north arrow, a scale bar, your name, and the date the map was created.

When you create a layout, ArcView automatically transfers the open document in the project to the layout. In this case, your open document is the view.

32 First, make any final adjustments to your map. When you're satisfied with the way it looks, click Layout from the View menu. Click Landscape and click OK.

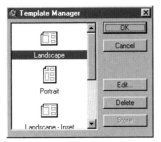

A layout is created for you. If necessary, adjust the size of your Layout window.

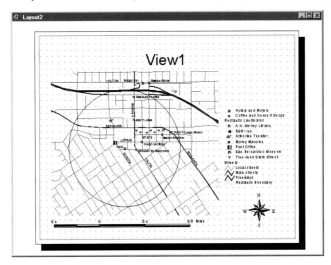

33 Click the Pointer tool and double-click the title text, View1. Change the title to **Downtown Redlands**. Click OK.

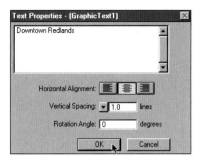

34 Click and drag Downtown Redlands so it's centered.

35 Click the legend and drag it up so you have room to put your name and date below it.

36 Click the Text tool. Position the cursor in the white space above the north arrow and click once. Type your name and date in the Text Properties window and click OK.

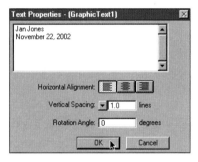

37 Click the Pointer tool and click and drag the name and date so they're centered between the map legend and the north arrow. You can also make the box bigger by clicking and dragging one of its handles. Resize and move other graphics until you are happy with the way the layout looks.

❊ **NOTE:** The layout often appears illegible and small on the computer screen. To check your work, you can zoom in on the layout. When you print the map, you will be able to read all the labels and symbols you worked so hard to create.

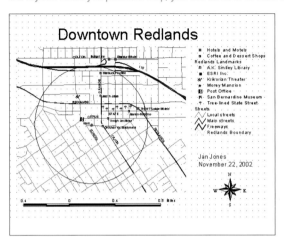

ACT

38 From the File menu, click Print. Click Setup to choose your printer settings (check Landscape and send to a color printer if one is available). Click OK to send the layout to the printer.

If you didn't like the way your map printed out, make adjustments to your layout and reprint.

39 Click the Project window title bar. Save your project. If you are in a classroom environment, ask your instructor for directions on how to rename the project and where to save it. For example, name your project **map_abc.apr** where "abc" represents your initials. Write down the new project name and location on a separate sheet of paper. Exit ArcView.

SUMMARY

In this exercise, you:

- Opened an ArcView project and added data to it
- Observed spatial relationships between themes and explored the attribute table
- Labeled features
- Created and printed a layout of Redlands, California, for visitors

---

MODULE 1 ACKNOWLEDGMENTS

Redlands data provided courtesy of the City of Redlands and ESRI and is used with permission.

All references to the Lincoln Memorial Shrine and A. K. Smiley Library are used with permission.

Reducing crime

In our imaginations, we can leave crime fighting to superheroes. Such characters can do almost anything—they can fly over cities to see where crimes are taking place and read the minds of criminals to guess where they will strike next. Superheroes don't need GIS, but real crime fighters do—many law enforcement agencies today use the power of GIS to fight and prevent crime. With GIS maps, police can see where crimes have occurred, and they can see patterns that help them understand crimes to solve them and to prevent future crimes.

CASE STUDY

Deciding where to increase neighborhood police patrols

The GIS students of Bishop Dunne Catholic School in Dallas, Texas, teamed up with the Dallas police department to help them establish a robbery task force. The robbery maps they produced at each stage of the project helped police justify the need for a task force, determine where and when to deploy task force patrols, and evaluate the patrols' effectiveness.

EXERCISE

Geocode crime data to map and analyze robbery hot spots

You will geocode Dallas robbery data, then classify it by attribute to map patterns of robberies of residences, businesses, and individuals at different times of day. You will identify the areas and times with the highest incidence of robbery and recommend where to carry out police task force patrols.

ON YOUR OWN

This section explores how to formulate partnerships between schools and local law enforcement agencies, such as police departments or neighborhood crime watch groups. Included are strategies for obtaining criminal data for your community, the best way to analyze the information, and how to communicate findings with the greater community.

Deciding where to increase neighborhood police patrols

Dallas, Texas

Robbery is an all-too-common crime in most large cities, and Dallas, Texas, is no exception. In the fall of 2001, the Dallas Police Department formed a fourteen-officer robbery task force to address a perceived robbery problem in the south-west part of the city. The task force would provide extra police patrols in trouble spots within the southwest division over a three-week period, with the hope that a greater police presence would increase arrests and reduce the robbery rate.

The challenge for the robbery task force was to find an effective way to assign patrol officers throughout the area. Where exactly were the trouble spots for robberies in the southwest division? What would be the most effective time of day for the patrols? Corporal James Allen, a community-policing officer for the southwest division and a task force member, thought the GIS students at Bishop Dunne Catholic School could help find the answers. In the end, students helped the police reduce robbery by nearly 40 percent in the robbery hot spots they identified through GIS analysis.

AN OPPORTUNITY FOR GIS STUDENTS

The GIS II class at Bishop Dunne began working with the Dallas Police Department in 1999 when they needed data for a community atlas project. Students wanted to generate a set of maps on aggravated assaults and robbery in the Oak Cliff section of southwest Dallas, where the school is located. At that time, Mr. Brad Baker, the class instructor, met Corporal Allen and the two developed an ongoing relationship to study and combat crime.

GIS instructor Brad Baker points to a robbery hot spot on a map of south-west Dallas as a student in his class and Corporal Allen look on. GIS students, guided by Corporal Allen's enthusiasm and dedication, began in 1999 to provide detailed crime maps each month to the twenty-four apartment managers in the Central Oak Cliff Crime Watch Zone. The project has since changed so that students now create a general monthly hot-spot map series for the southwest division commanders.

When Corporal Allen approached the class with a new task in October 2001, students were eager to participate. He asked them to provide intelligence for the newly formed robbery task force. Familiar with the quality of the students' work, Corporal Allen felt that the students might both improve the results of the task force and increase interaction between the police and community.

Mr. Baker's first task was to gain approval for students to work on the project. Because of the successful 1999 mapping project and the established relationship with Mr. Baker, police department approval was granted quickly.

ASKING QUESTIONS, FINDING ANSWERS

For the next seven weeks, six students in grades 10 through 12 served as crime analysts, providing intelligence for the robbery task force. Answering the apparently simple questions of where and when to assign task force patrols required accomplishing a series of complex analytical tasks using GIS. In the end, students found themselves going through the geographic inquiry method multiple times. There was no one right answer, nor one way to look at the data. Students ultimately created six series of maps to answer a variety of questions.

Before detailed analysis could begin, the team had to justify that there was a definite need for the task force. Students needed to create a map that would answer the question, "Does the Dallas Police Department's southwest division have a problem with robbery?" They envisioned a map of the entire city showing the pattern of robberies for August, September, and October.

The class requested the data from their liaison in the police department, who removed any confidential and sensitive information before supplying it to the students. The robbery data began in the form of a table containing the address location and other attributes for each crime. To map the robberies, the addresses in the table would need to be geocoded, or matched, to those on a street data layer. Students were fully trained in geocoding procedures, but the decision was made to have the police department geocode the data. This decision allowed students to meet the rapid time line that was required for the project and use the most accurate street data possible from the department's GIS database. The students received ArcView shapefiles containing a series of robbery points.

After mapping the robbery points in ArcView, the Bishop Dunne students decided to further analyze the data using raster analysis. (A raster is a data set made up of a grid of equally sized square cells. Each cell has a value representing a measurement or attribute of the area the cell represents.) With raster analysis, students could create a continuous robbery "surface" across the entire area. And, they could quickly evaluate changes over time by subtracting cell values in two different rasters and producing a "difference map."

Students used the ArcView Spatial Analyst extension to create the continuous raster, interpolating between the discrete robbery points and calculating cell values that reflected the density of robberies in a particular area.

The next phase was to help the task force define the most effective patrol areas and times. Students generated maps showing robbery patterns at different times of day (watches) as well as areas with the greatest concentration of robberies (hot spots). Ultimately, the police task force defined ten new patrol

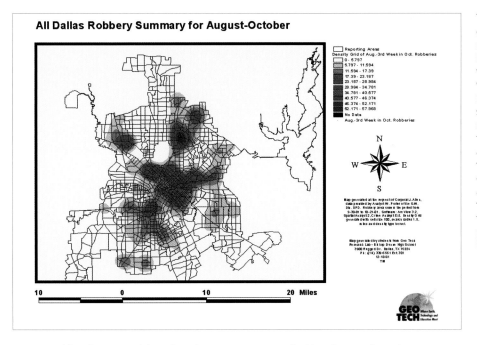

The first map students generated showed that the recent pattern of robberies in the southwest division was less intense but more widespread than in the rest of the city. The map shows data from the three months prior to the task force (August, September, and October). It depicts the robbery raster with outlines of police reporting areas. Based on the map, the students recommended that the robbery task force proceed with planning immediately.

zones, taking into consideration the patterns revealed by the students' maps as well as their own extensive experience with criminal activity in the area.

The task force began its patrols, which continued for three weeks. To be most effective, the police officers needed current information about robbery activity within individual patrol areas. Students decided that they could best support the officers by generating daily patrol zone maps showing detailed street locations of the latest robberies.

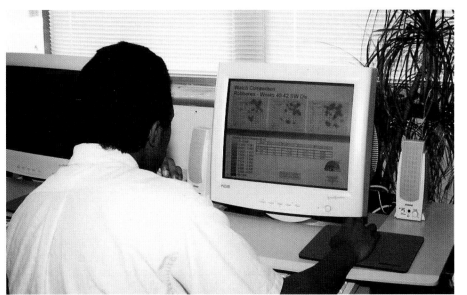

A student in the Bishop Dunne GeoTech computer lab compares three robbery maps of southwest Dallas. Each map shows the robbery pattern for a different police watch.

Module 2   Reducing crime

21

case study

The students determined the best time frame for the patrols by mapping the data according to three different watches. From the watch comparison maps and summary table, students could see that Watch 3, from 4:00 P.M. to midnight, was the time of greatest concern. Students were surprised to see that the most intense hot spots appear to move with each watch. The decision was made to conduct the robbery task force patrols from 4:00 P.M. to 2:00 A.M. each day.

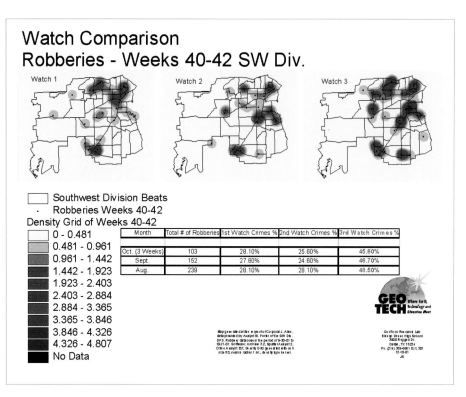

Watch Comparison
Robberies - Weeks 40-42 SW Div.

Watch 1   Watch 2   Watch 3

☐ Southwest Division Beats
· Robberies Weeks 40-42
Density Grid of Weeks 40-42
- 0 - 0.481
- 0.481 - 0.961
- 0.961 - 1.442
- 1.442 - 1.923
- 1.923 - 2.403
- 2.403 - 2.884
- 2.884 - 3.365
- 3.365 - 3.846
- 3.846 - 4.326
- 4.326 - 4.807
- No Data

Month	Total # of Robberies	1st Watch Crimes %	2nd Watch Crimes %	3rd Watch Crimes %
Oct. (3 Weeks)	103	28.10%	25.60%	45.60%
Sept.	152	27.60%	24.60%	46.70%
Aug.	239	28.10%	28.10%	48.50%

To find targets for task force patrols, students analyzed the robberies that occurred in the last three weeks. They used ArcView Spatial Analyst and Crime Analysis Application extensions to create density rasters showing the greatest probability for recurrent crime. Their analysis revealed five such areas, outlined in red on the map. Police reporting areas (subareas of beats) intersecting with these hot spots are shown in yellow. Police decided to focus their efforts on three of the five hot spots.

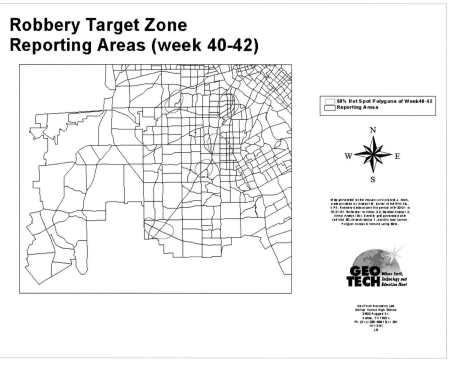

Robbery Target Zone
Reporting Areas (week 40-42)

☐ 50% Hot Spot Polygons of Week40-42
☐ Reporting Areas

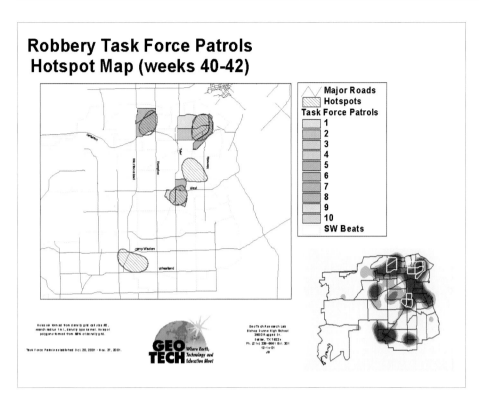

After the task force defined ten task force patrol areas, students created this map to show the Southwest Command the patrol locations in relation to the five robbery hot spots. Four patrol zones were placed in areas outside the hot spot because the police felt it would be important to patrol neighboring residential areas where offenders might go immediately after a robbery. On the large map, major streets appear in blue. The inset map shows the patrols (white) in relation to the robbery pattern for the entire division.

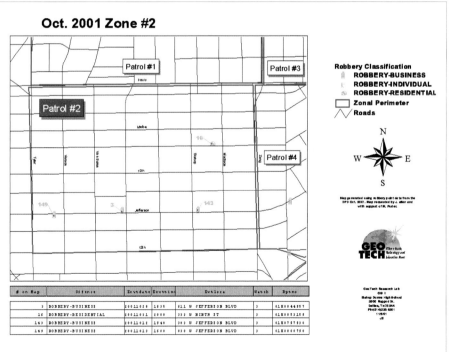

Where did robberies occur recently? Twice during the project, students generated maps like this one to help officers see the individual locations of robberies affecting businesses, residences, and individuals in individual patrol zones. Information on each incident is listed in the table below the map.

MEASURING SUCCESS

Just how effective was the task force in stopping robbery? To find out, students put the GIS to work once again. A comparison of the crime statistics before, during, and after the robbery task force patrols showed a dramatic decrease in robbery in the areas of the task force patrols. Even though robbery in other areas of the southwest division stayed the same or increased, the overall result was a reduction in crime for the entire division. Both the students and the police learned that the task force alone could not wipe out crime in an area. A more enduring reduction in crime requires a cooperative community effort through a variety of programs.

Students created robbery density rasters for each three-week period before, during, and after the task force patrols. They subtracted one raster from another using the ArcView Spatial Analyst map calculator to see where crime increased or decreased during and after the patrols. The bar graph shows that during the patrol period, robbery dropped a dramatic 39 percent in the patrol areas and 19 percent over the entire southwest division. In the following weeks, the decrease continued in the patrol areas, probably because the task force led to the arrest of as many as three perpetrators. However, robberies did rise quickly for the division once the patrols stopped.

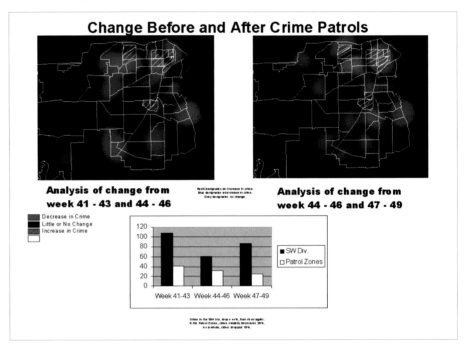

Change Before and After Crime Patrols

Analysis of change from week 41 - 43 and 44 - 46

Analysis of change from week 44 - 46 and 47 - 49

NEXT STEPS

The students of Bishop Dunne Catholic School will continue to work closely with Corporal Allen and the Dallas Police Department. Students continue to generate monthly robbery hot-spot maps. They also plan to begin mapping serial offenders, pairing up GIS student groups with detectives on specific cases. Together, the GIS students and police will continue to establish renewed community policing programs in the neighborhood near the school. Through their continuing partnership, the students have found a unique way to make a difference in their local community.

SUMMARY

ASK A GEOGRAPHIC QUESTION	• Does southwest Dallas have a problem with robbery? • Where are robberies occurring? • When are robberies occurring? • Where and when should task force patrols take place? • Where are the new patrol areas in reference to the robbery hot spots? • Where are crimes occurring within the patrol areas? • How effective was the task force in stopping robbery? • What areas should be targeted next?

ACQUIRE GEOGRAPHIC RESOURCES	• Robbery data for:
	  – Three months prior to task force
	  – Three weeks of task force
	  – Three weeks after task force
	• Local street data for address geocoding
	• Police patrol areas
	• Major streets and freeways

EXPLORE GEOGRAPHIC DATA	• Geocode robberies to create points • Calculate robbery surface rasters • Digitize hot-spot areas • Digitize task force patrol zones • Create daily robbery maps with reference maps
ANALYZE GEOGRAPHIC INFORMATION	• Discover and describe spatial and temporal robbery patterns across the city and in the southwest division
ACT ON GEOGRAPHIC KNOWLEDGE	Make recommendations to: • Proceed with task force • Patrol from 4 P.M. to 2 A.M. • Create ten patrols covering three of the five hot spots • Generate individual patrol zone maps detailing recent robberies • Include results in future antirobbery initiatives

Geocode crime data to map and analyze robbery hot spots

When police officers prepare a crime report, they record the street address where the crime took place. In order to analyze crimes geographically, the addresses need to be matched to their locations on a street map. This process is known as geocoding.

Reference data

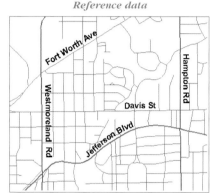

Address table

Address	Date	Time	Offense
2700 W DAVIS ST	10/15/2001	830	ROBBERY-INDIVIDUAL
3300 W DAVIS ST	10/18/2001	2235	ROBBERY-INDIVIDUAL
2427 W JEFFERSON BLVD	10/8/2001	400	ROBBERY-INDIVIDUAL
2515 W JEFFERSON BLVD	10/11/2001	1615	ROBBERY-BUSINESS
2871 FORT WORTH AV	10/13/2001	1930	ROBBERY-INDIVIDUAL

   In this exercise you will geocode some of the Dallas robbery data analyzed by the students of Bishop Dunne Catholic School. Once you have mapped the robbery locations, you will analyze the patterns made by robberies and identify and examine areas where robberies seem to be clustered. These areas are called hot spots.

The ✉ icon indicates questions to be answered. Write your answers on a separate sheet of paper.

ASK

You have been asked to assist the Dallas Police Department by identifying robbery hot spots within the southwest division during the three-week period from October 7 to October 27, 2001. You are to submit a report in which you identify the five most prominent hot spots, and compare and contrast them according to the number and types of robbery that occurred in each hot spot. Finally, you must select five beats within the hot spots that you believe could benefit from extra police patrols. A beat is an area that police patrol. The police will use the information to decide where to assign their robbery task force patrols.

ACQUIRE

You will use GIS to map and analyze the robbery data. The department's crime analysis unit has provided you with the following data:

DATA SET DESCRIPTION	DATA FORMAT	FEATURE TYPE	FILE NAME
Roads	Shapefile	Polyline	SW_roads.shp
Major arterials	Shapefile	Polyline	SW_arterials.shp
Southwest division police beats and zones	Shapefile	Polygon	SW_beats.shp
Robbery data for October 7–27	Text file	—	Crime.txt

PART 1   EXPLORE AND GEOCODE ROBBERY DATA

EXPLORE

1 Start ArcView. Click the File menu and choose Open Project. Navigate to the exercise data folder (*C:\esri\comgeo\module2*) and open *crime.apr.*

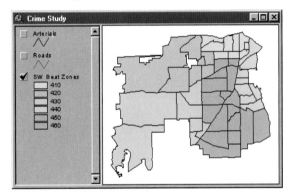

Individual police beats in southwest Dallas are outlined in black on the map. They are shaded different colors to indicate the various beat zones (zones 410 to 460). Two additional themes, Arterials and Roads, are listed in the table of contents.

2 Click the check boxes for the Arterials and Roads themes to display them.

At this map scale, the roads look crowded. Even so, you can see that the overall pattern of the road network is denser on the right side of the beat map than on the left side. To get a better look at the roads, you'll zoom in on them.

3 Click the Zoom In tool. Drag a box in the upper right corner of the beat map.

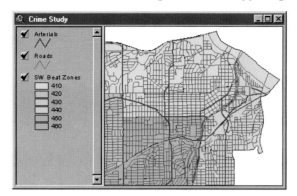

You will be using the Roads theme for geocoding, so it's a good idea to find out what fields are contained in this theme. A theme used in geocoding needs to have fields having address information, such as street names, address ranges, or ZIP Codes.

4 Click the Roads theme in the table of contents to make it active. Click the Identify tool, then click one of the gray roads on the map.

※ NOTE: Several roads may be listed in the Identify Results window because they are close together on the map.

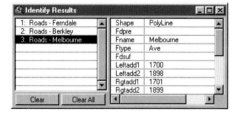

5 Scroll down the list at the right and look at the fields and the information they contain. To see the information for a different road, click another one in the list at the left.

5a Which field contains the road's name?

5b Which field tells you if the road is called a street, avenue, lane, or so on?

5c How many fields in the list contain ZIP Code information?

6 Close the Identify Results window. Click the Zoom to Previous Extent button to return to the full extent of the southwest division.

The process of geocoding has three basic steps. You need to (1) make a theme *matchable,* (2) match a list of addresses, and then (3) correct any unmatched addresses. To make a theme matchable, you need to specify which fields contain the address information and which address style should be used; ArcView uses this information to build the *geocoding index.* Once the index is built, the theme is matchable. In this project, the Roads theme has the address information you need for geocoding, so you will make this theme matchable.

7 **Make sure the Roads theme is active. Click the Theme Properties button. On the left side of the Theme Properties window, click Geocoding. Set the Address Style to US Streets (not US Streets with Zones).**

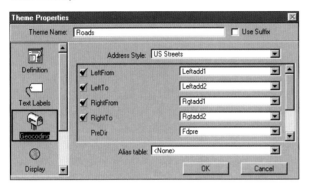

8 **Scroll down the field list.**

Notice which Roads fields ArcView has automatically selected to get the information it needs for geocoding. The fields required for the US Streets style are indicated with check marks.

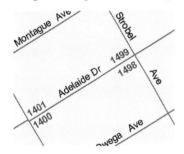

In the US Streets style, address numbers on individual street segments are represented using four attribute fields: LeftFrom, LeftTo, RightFrom, and RightTo. In this example, 1400 is the RightFrom address (the lowest number on the right-hand side as you would face the higher numbers). At the far end of the block, 1498 is the RightTo address. In between, the other right-side addresses are assumed to be even numbers. The odd numbered addresses are on the left side of the street, ranging from 1401 to 1499.

 8a Which five pieces of information must ArcView have to geocode according to the US Streets style?

9 Click OK. A message appears asking if you want to build geocoding indexes using the address style you selected. Click Yes.

You can watch the progress in the ArcView status bar as ArcView takes a moment to prepare the index for the Roads theme. The Theme Properties window closes when the index is finished.

It's a good idea to do a test with a single address before you attempt to geocode the long list of crime addresses. You will test the system using the address of Bishop Dunne Catholic School.

10 Click the Locate Address button. In the Locate Address box, type **3900 Rugged Dr.** and then click OK.

10a What happens?

✳ **NOTE:** The locator dot is only a graphic drawn on the map and does not have any data associated with it. When you geocode, a new shapefile will be created. The locations will be saved as points in this new theme, and each point will have robbery information associated with it.

Your test was a success! You don't need to mark this location anymore, so you will delete the dot.

11 Click the Pointer tool, then click the dot to select it. Press the Delete key on your keyboard.

Now that the Roads theme is matchable, you are ready to locate the robberies. The robbery data is contained in a text file. In order to work with it, you need to add it to the project as a table.

12 Bring the Project window *(crime.apr)* to the front. Click the Tables icon, then click Add. If necessary, navigate to the exercise data folder *(C:\esri\comgeo\module2).* Change "List Files of Type" to Delimited Text (*.txt), and add *crime.txt.*

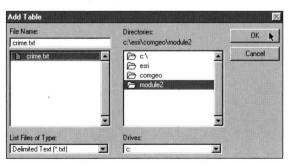

The crime.txt table is added to the project and displayed.

13 Examine the crime.txt table.

13a How many crimes are recorded in the table?

⁘ **HINT:** Look at the left side of the toolbar.

13b On a separate sheet of paper, make a table like the one below. Indicate what information is contained in the table by writing the field names (or "not included").

INFORMATION	FIELD NAME
Address where the crime took place	
ZIP Code where the crime took place	
Type of crime	
Watch	
Time of day	
Police officer on duty	
Beat number	

14 Close the crime.txt table.

15 Make the View window active, then click the View menu and choose Geocode Addresses.

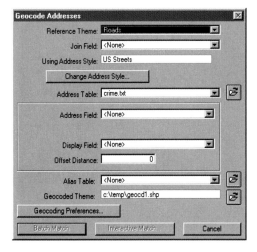

Check the Geocode Addresses window to make sure the correct reference theme (Roads), address style (US Streets), and address table (crime.txt) are selected and make any necessary changes.

Next, you will select the address field contained in crime.txt and specify a distance of one-tenth mile to offset the geocoded point from the street. Offsetting a point allows you to see what side of the street the address is on.

16 Click the Address Field drop-down list and choose Evtloca. Enter **0.1** in the Offset Distance box.

Finally, you will choose a name and location for the crime shapefile that will be created.

17 Click the file button next to the Geocoded Theme box. In the Geocoded Theme Name window, navigate to the exercise data folder. If you are in a classroom environment, ask your instructor where to save your work for this exercise. In the File Name box, type **crime** with an underscore and your initials, for example **crime_abc.shp**. Click OK.

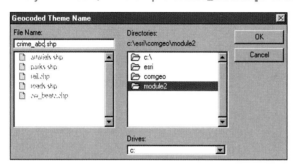

18 Make sure your Geocode Addresses window matches the following graphic, then click Batch Match to start address matching.

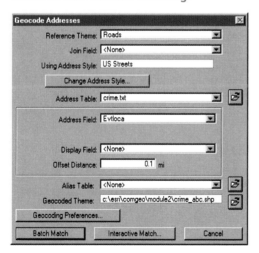

ArcView compares the addresses in the crime table with the address information in the Roads table. This may take a few moments. When it's finished, the Re-match Addresses window appears.

19 Examine the information in the Re-match Addresses window.

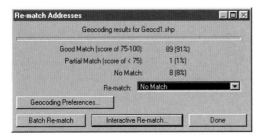

🖎 19a How many addresses found a good match?

🖎 19b What percentage of the addresses is this?

Notice that one address had a partial match, and eight had no match in the Roads theme.

Partially matched or unmatched addresses can happen for a variety of reasons. For example, a road name could be misspelled in the crime table, the roads table, or both. In many cases, you can identify and fix the problem that allows you to look at each situation individually.

20 Click the Interactive Re-match button.

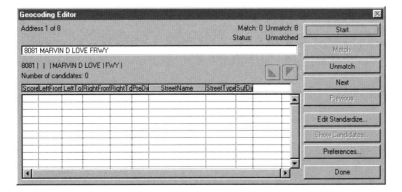

The first unmatched address, 8081 MARVIN D LOVE FRWY, is listed at the top of the Geocoding Editor window. ArcView has not found any similar addresses, or candidates, for this record. However, if you were to do a little research in the Roads theme table, you would find a road named MARVIN D LOVE SERVICE ROAD. A service road generally runs parallel to a highway, and businesses or homes actually are located on the service road rather than on the highway itself. Thus, you will assume the service road is the correct road.

21 In the Geocoding Editor window, delete the word FRWY from the address and type **SERVICE ROAD**. Press the Enter key.

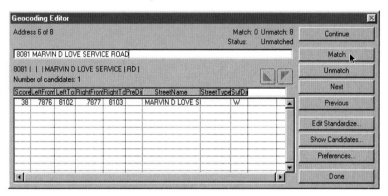

A possible match now appears in the candidate list. (It's important to note that the typing you did in the address box helped ArcView find the candidate, but it did not change the original data in the crime.txt table.)

22 Compare the address with the characteristics of the candidate.

22a Is the building number (8081) within the address ranges for this street segment?

22b Does the street name match exactly?

22c What information does the candidate include that is not contained in the address?

Although this address still has a low score (38), you are confident this is the correct street segment, so you will create a match.

23 Click the Match button.

The next address is NINTH & STARR, which is an intersection between two streets. This time a candidate is listed. The addresses don't match exactly because the Roads table uses "9th" instead of "Ninth," but again you are confident it is a match.

24 Click the Match button again. Continue checking and matching the remainder of the addresses. Look for discrepancies such as different spellings, lack of a directional notation (N, S, E, W), or missing spaces. Assume for the purposes of this exercise that all the candidate addresses are correct matches. When you finish, all the addresses should be matched.

25 Click Done in the Geocoding Editor window, then click Done in the Re-match Address window.

26 In the Crime Study view, turn off the Roads theme. Turn on the new Crime_abc.shp theme. Open the Legend Editor for the Crime_abc.shp theme. Change the symbol to an outlined circle, and change the color to bright blue so that the dots are easier to see.

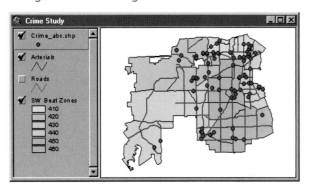

26a What patterns do you see in the distribution of robberies?

26b Do most robberies occur along major streets (arterials) or away from them?

27 Make sure the Crime_abc.shp theme is active and open its attribute table. Examine the fields in the table, scrolling all the way to the right.

Av add	Av status	Av score	Av side
400 N LANCASTER AV	M	100	R
500 N GILPIN AV	M	77	R
8081 MARVIN D LOVE SERVI	M	38	R
3601 S MARSALIS AV	M	100	R
3287 S POLK ST	M	100	R
712 S WALTON WALKER BL	M	100	R
NINTH & STARR	M	55	
225 S LANCASTER AV	M	100	R
418 N GILPIN AV	M	77	R
5000 S COCKRELL HILL RD	M	75	L
3900 SUMMIT RIDGE DR	M	100	L
800 N LANCASTER AV	M	100	R
2703 ALDEN AV	M	100	R

Do the fields look familiar? The fields in the crime.txt table were copied into this one during the geocoding process. The fields at the end of the table beginning with AV_ were added by ArcView. These four fields tell you the address used, whether the address had a match, the score, and whether the address is on the right or left side of the street.

28 Close the table.

29 Click the Project window title bar. Save your project. If you are in a classroom environment, ask your instructor for directions on how to rename the project and where to save it. For example, name your project **crime_abc.apr** where "abc" represents your initials. Write down the name and where it is stored. Exit ArcView if you are not continuing to part 2 until later.

PART 2   ANALYZE ROBBERY PATTERNS

ANALYZE

You'll begin your data analysis by visually analyzing the map for clusters of robberies. You'll draw circles around the five most prominent clusters and label these hot spots 1 through 5 on the map.

1   If necessary, start ArcView and open the crime project that you saved at the end of the first part of the exercise.

First you'll choose a more descriptive name for the crime theme.

2   Open the Theme Properties window for the Crime_abc.shp theme in the Crime Study view. Type **All Robberies** in the Theme Name box. Click OK.

3   Turn off all the themes except All Robberies.

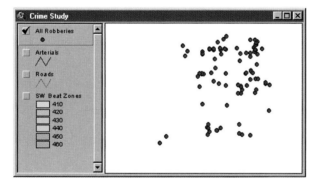

Notice that some of the robbery symbols are clustered and others are not. Next you will create a new shapefile showing where clusters exist (hot spots) and set up the symbol to use for them.

4   From the View menu, choose New Theme. Select Polygon as the feature type in the New Theme dialog and click OK.

5 Navigate to the location where you are saving your work for this exercise. Name the file **hotspots_abc.shp** where "abc" represents your initials. Click OK.

Hotspots_abc.shp is added to the table of contents. The dotted box around the theme's check box indicates that the theme is ready for editing.

Next you will change the symbol so that the hot spots will be semitransparent with a thick outline.

6 Open the Legend Editor for the Hotspots_abc.shp theme. Double-click the symbol. In the Fill Palette, click the box in the first column, third row. Change the outline size to 3.

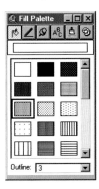

7 In the Color Palette, make the Foreground color red, the Background color none (the ✕ in the first box), and the Outline color red. Click Apply in the Legend Editor. Close the Legend Editor and Symbol Palette windows.

8 Click and hold the Draw tool to display the menu of drawing tools. Select the Draw Circle tool.

9 Locate a cluster of points on the map that you think is a robbery hot spot. Place the crosshair of the Draw Circle tool in the center of the cluster. Click and drag the mouse to draw a circle around it. To adjust the circle once it is drawn, use the Pointer tool to move or resize it.

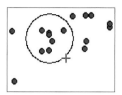

10 Locate four more clusters and draw circles around them so that you have five hot spots in all. You may want to zoom in a little.

✳ **NOTE:** You may observe more than five clusters of points on the map that you could circle. Remember, the Dallas Police Department has asked you to focus your analysis on the five most prominent hot spots. As the crime analyst, you must decide which five clusters are the most important and precisely which points they should include.

11 When you are satisfied with your five circles, click the Theme menu and click Stop Editing. Click Yes to save your edits. Clear any selected features, if necessary.

12 Click the Zoom to Active Theme button so you can see all five hot spots. Turn on the SW Beat Zones theme.

Next you will number the hot spots.

13 From the Window menu, choose Show Symbol Window. In the Text Palette, choose Arial Black, size 18. In the Color Palette, set the text color to red. Click the Text tool and number each circle from 1 to 5. Place the number just outside the circle.

✳ **NOTE:** If you are working with ArcView 3.1 or higher, you could use the text callout tool instead of plain text.

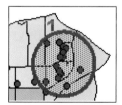

14 From the Edit menu, choose Select All Graphics. Then from the Graphics menu, choose Attach Graphics. Attaching these graphics to the Hotspots_abc.shp theme will allow you to turn them on or off with the hot spots. Click the Pointer tool and click on the map away from the numbers to clear the selection.

✳ **NOTE:** Remember to save your project periodically! This would be a good time to do so.

Now that you have identified robbery hot spots based on your visual observation of the data, it's time to take a closer look at what's going on in those areas. You will classify the robberies according to type and use special crime symbols to display them.

15 Make the All Robberies theme active. From the Edit menu, copy and then paste the theme so you have a duplicate.

16 Change the name of the duplicate theme to **All Robberies by Type**. Refer to part 2, step 2 if you need help.

17 Open the Legend Editor for the All Robberies by Type theme. Change the Legend Type to Unique Value. For the Values Field, choose Offense.

Different colored symbols are used for business, individual, and residential robberies. The long labels will be difficult to read in the table of contents. You will make them shorter.

18 In the Label field, delete "ROBBERY-" from the front of each label.

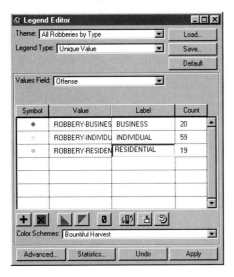

19 Click Apply. Move the Legend Editor so you can see the map. If necessary, turn layers on or off to answer the following question.

19a How does the pattern of business robberies in the southwest division differ from that of robberies of individuals?

To make the different types of robberies stand out even more, you will change the symbols from dots to pictorial markers. ArcView comes with many different symbols, but a limited selection is loaded by default. You will add the ESRI Crime Analysis markers to the palette.

✳ **IMPORTANT NOTE TO MACINTOSH USERS:** Before proceeding to step 20, you will need to copy the ESRI Crime Analysis font suitcase from the exercise data folder (comgeo:module2:ESRI Crime Analysis.suit) to your computer's Fonts folder in the System Folder. You will need to save the project, then exit and restart ArcView for the font to be recognized.

20 In the Legend Editor, double-click the symbol for business robberies to open the Symbol Palette. Click the Font Palette button, then scroll down the list and click ESRI Crime Analysis. Change the size to **14**.

21 Click the Create Markers button.

Click the Marker Palette button if necessary. Scroll down the list of markers to find the new crime symbols. They start with the following symbols:

22 Choose a marker symbol that you think best fits each type of robbery. Make each symbol size **14**. If you choose, you can change the symbol color as well.

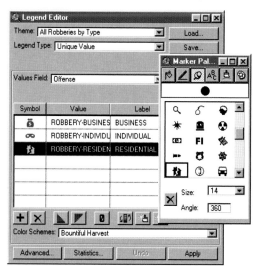

23 Click Apply in the Legend Editor to update the map.

You will save this legend in case you want to use it again later.

24 Click the Save button in the top right of the Legend Editor. In the Save Legend dialog, save the legend as **offense.avl** in your designated data directory and click OK. Close the Legend Editor and Symbol Palette windows.

Next, you'll analyze the types of robbery within the hot spots.

25 Make sure the All Robberies by Type and Hotspots_abc.shp themes are turned on. The All Robberies theme should be turned off so it doesn't visually interfere with the robbery type symbols. If necessary, zoom and pan the map to examine the hot spots and answer the following question.

25a In general, are certain types of robbery more prevalent in some hot spots than in others? Describe your observations.

26 Turn on the Roads and Arterials themes.

26a What is the relationship of robberies to arterial streets in the hot spots you identified? For example, does a string of similar robberies occur along a particular arterial?

Next, you will use the information in the theme's attribute table to calculate the number of each type of robbery as well as the total number of robberies in each hot spot.

27 Turn off the Roads and Arterials themes and turn on the SW Beat Zones theme. Zoom the map so you can see all five hot spots.

28 Press Shift and make both of the following themes active: Hotspots_abc.shp and All Robberies by Type.

29 Click the Select Feature tool and select hot spot 1. (When it is selected, the shading within the circle turns yellow but the outline does not.)

30 From the Theme menu, choose Select by Theme. In the dialog, you need to construct a query that reads "Select features of active themes that **Are completely within** the selected features of **Hotspots_abc.shp**."

※ NOTE: You must select from the bottom list before you will see the correct choice in the top list.

Click New Set. All the robberies in hot spot 1 are selected.

You will take a closer look at the attributes of the selected robberies.

31 Open the All Robberies by Type attribute table. Click the Beat field name and click the Sort Ascending button.

32 Click the Promote button to bring the selected records to the top of the table.

32a Make a table like the one below. In the second column, first row of your results table, write the ID numbers of the police beats in which the selected robberies took place.

Hot spot results table

HOT SPOT	IDs OF BEATS IN HOT SPOT	BUSINESS ROBBERIES	INDIVIDUAL ROBBERIES	RESIDENTIAL ROBBERIES	TOTAL ROBBERIES
1					
2					
3					
4					
5					

To fill out the rest of the information for hot spot 1, you will have ArcView summarize the number of robberies by type of offense, rather than count each record yourself.

33 Click the Offense field name in the attribute table. Click the Summarize button. Click OK in the Summary Table Definition dialog to generate a simple summary table showing a count of robberies by type (e.g., sum1.dbf).

33a Record your findings in your results table. Add up the total robberies yourself to complete the last column.

34 Close the summary table. Bring the View window to the front and click the Clear Selected Features button.

34a Repeat steps 28–33 for each hot spot and record your findings.

Now you have a better understanding of the number and types of crimes in each hot spot. However, the police will need to assign their task force patrols by beat.

ACT

35 Study the observations you recorded above. Zoom in on the areas that you feel most need attention, considering the hot spots. Use the Identify tool to help you recommend five beats where the Robbery Task Force should focus special patrols.

35a List the type of crime police should focus on while patrolling the beat.

36 Save your project.

SUMMARY

In this exercise, you:

- Geocoded tabular robbery data and displayed the robberies as points on a map
- Drew circles around clusters of robbery points and added them to a shapefile
- Calculated summary statistics for a set of selected robberies
- Compared and contrasted robbery hot spots by interpreting the summary statistics
- Recommended five locations for increased police patrols

ASK A GEOGRAPHIC QUESTION

We usually become aware of crime in our communities one incident at a time. The evening news reports a murder, for example, or a classmate relates that his bicycle was stolen. Perhaps you have even witnessed a crime or been a victim. When several notable incidents happen close to each other or within a short period of time, we begin perceiving patterns of crime in space or time. Think of your own community for a moment—do you think of particular neighborhoods or sections of town as "high-crime" areas and others as "safe"?

Usually, our perceptions do not tell a complete or accurate story of what is really going on with crime in a community. Even the perceptions of a police officer with a lifetime of experience in a community can be incomplete. Mapping crimes and finding geographic patterns can provide insights that in turn may lead to better crime-fighting solutions. Maps and geographic analysis, together with first-hand experience and knowledge of a community, allow people to find more effective ways to deal with crime, from catching criminals to identifying which neighborhoods or schools to target with innovative crime-prevention programs.

Choose a crime-related issue to focus on

Crime is a broad, complex subject that covers a wide spectrum of events and causes. Your most important task at this point is to narrow the focus of your project. Choose a crime-related issue that is currently relevant to your community and to you personally. In Dallas, students analyzed robbery data, but you may find that reducing litter or graffiti is a more relevant issue for you. Try to choose an issue that has a clear locational component.

Once you have selected a possible topic, make a list of geographic questions related to that topic. For example, you may ask, "Where are crimes taking place? How does the rate of crime in one area compare with another? What are the common characteristics of the places where a particular type of crime is common (or uncommon)? Are there neighborhoods where the population is especially vulnerable to crime (e.g., senior citizens, young children, immigrants)? How do the patterns of crime change over time?" Then consider the amount of time you have for your project. You may need to narrow your focus even further by choosing just one or two questions to begin with.

The following suggestions will help you pose and select geographic questions to investigate.

- Brainstorm crime issues that your community is currently facing. Look at the headlines of your local newspaper to help you with ideas.

- Think of people in your community who may have a shared interest in the topic. Local law enforcement officials are an obvious example, but don't overlook such potentially useful partners as crime-watch groups, neighborhood improvement groups, or youth centers. Discuss with them questions that need to be answered or problems that need to be solved.

What crime-related issues are important to your community?

You can use geocoding skills to map and understand:

- ➤ Graffiti or vandalism locations
- ➤ Illegal trash-dumping sites
- ➤ Automobile thefts
- ➤ Crimes around schools or playgrounds
- ➤ Properties damaged by arson
- ➤ Drug-related arrests
- ➤ Vacant lots

- Choose a question whose time sensitivity matches the schedule and amount of time you have available to work on the project. For example, police departments need some kinds of information on a daily, or even minute-by-minute, basis. Volunteering to meet such information needs is probably not realistic. Questions about long-term trends, before-and-after analysis, or what-if scenarios are less time-critical and therefore may be more appropriate for a community project.

- Narrow the focus of your question. In the case study, for example, the focus was narrowed from "crime" to "robberies," from "all of Dallas" to "the southwest division," and from "three months" to "three-week periods before, during, and after the task force patrols."

ACQUIRE GEOGRAPHIC RESOURCES

Here are some types of data you may need for your crime investigation and possible sources of such data.

TYPE OF DATA	POSSIBLE DATA SOURCES
Crime statistics	• Your local police department's Open Records Unit • Weekly police blotters published in the local newspaper
Street data for geocoding	• Your community's GIS or city planning department • Your state or county Web site • GIS users in your community (e.g., utility companies, phone companies) • Data sites on the Internet such as U.S. Census TIGER/Line data from *www.geographynetwork.com* (see the Community Geography Web site *www.esri.com/communitygeography* for additional sources)
Other geographic features needed for reference or analysis, such as parks, schools, shopping centers, rivers, and so on	• Your community's GIS or city planning department • State natural resource information office (often available on the Web) • Regional council of governments Web site • U.S. Geological Survey (USGS) for topographic maps and aerial photographs on *terraserver.Microsoft.com* and other data via *earthexplore.usgs.gov* • USGS and other basemapping data via *www.gisdatadepot.com*

Tips

- You may want to contact your local police department and plan to visit them. Larger cities usually have several community contact people who can assist you in acquiring information such as the name of a community policing officer. In addition, many cities have a GIS department in place, and you may be able to obtain data that is already in a form your GIS can read (e.g., shapefile format). If such a department exists, contact its crime analyst or GIS specialist for assistance.

- If you live in a smaller town that does not have GIS data, check to see whether it is available from a larger entity such as a county or state agency or a regional intergovernmental association. Besides providing data, a contact at such an agency may also help you better frame your geographic question and organize your study of the data.

- Crime data, like other kinds of data, comes in a variety of forms that you can use in GIS software, from tab-delimited text files (like the one used in the exercise) to files in a particular database format. Find out from your source which data formats they can provide so you can request the one that's most convenient for use with your GIS software. You should also consider the impact the data format might have on how you will analyze the data.

- Uniform Crime Reporting (UCR) standards are used by law enforcement agencies across the country. These standards for reporting murder, robbery, aggravated assault, motor vehicle theft, and arson, among other crimes, allow police departments to communicate criminal information effectively no matter where the crime occurred and to solve crimes more efficiently. Visit the Community Geography Web site at *www.esri.com/communitygeography* to find links to information on UCR standards.

- In the United States, crime data can be obtained through the Open Records provision of the Freedom of Information Act. Refer to "On your own: Project planning," page 251 and the Community Geography Web site for more on this law and how to submit an Open Records request.

- Be aware of privacy issues. Although criminal activity is considered part of the public record in the United States and is often published in summary form in newspapers, some police records contain private or sensitive information. This could be information such as victims' names or details that must remain private because of ongoing investigations. Check with your local law enforcement agency to identify potential privacy issues.

- When acquiring potentially sensitive crime data, take special care to obtain metadata that states any use constraints, liability disclaimers, special processing or codes designed to ensure privacy, and the data source. Be sure you are clear about whether your planned GIS activities and community actions (e.g., mapping, putting maps on a Web site, or sharing maps with the public) are permitted uses of this data.

- The right street data is essential to successful geocoding. (See also the "Technical issues" section under "Explore geographic data" below.) Be sure that the street data you acquire for geocoding has the following characteristics:

  - Includes matchable address fields. Street address fields should be in a form that can be matched to your crime data. Generally, your street data should include address ranges, street names, street types, and directional information (i.e., N, S, E, W).

  - Is as error-free as possible. This means that the data has relatively few errors such as street name misspellings, missing or misspelled street types, or missing directional information such as E (east) or W (west).

  - Is as current as possible. Many areas of the country experience frequent additions of new suburban developments, and you'll probably want these new streets to be included on your map.

- Remember that data other than crime locations can be geocoded. Grocery stores, movie theaters, hospitals, banks, and so on can be easily mapped if you have a list of addresses, for example from a telephone directory. Depending on your crime issue, you may want to include features like these on your map.

- Think of people who are "experts" who may be able to help you frame analysis or provide insight into the patterns you uncover. For example, you could interview long-time residents of your neighborhood about their perceptions or observations of crime in the area you are studying. Don't overlook professionals like lawyers, detectives, social workers, city council members, and others who may have knowledge about or interest in the crime you are investigating.

EXPLORE GEOGRAPHIC DATA

Technical issues

The first step in exploring crime data typically is to geocode it so you can explore it geographically. Because the success of your geocoding depends on a number of variables (mostly related to the quality of your data sources), it sometimes can be a tedious or frustrating process. You may want to read up on geocoding to help you better understand how to avoid potential problems (or solve those you've already encountered). Visit the Community Geography Web site for a list and links to some of the many resources on the subject.

Preliminary data exploration

Before beginning your actual analysis of the crime data, take time to explore it and familiarize yourself with its content, characteristics, and potential spatial patterns.

Here are some suggestions for preliminary exploration:

- Look for general patterns in the way crimes are distributed in your community. What are some possible explanations for the patterns you observe? For example, if crimes are clustered in a particular area, is there a geographic feature such as a business or shopping center that could be a magnet for the type of crime you are examining?

- Explore the crimes in a variety of ways. For example, create a unique-value map for type of criminal activity, time of day, time of year, and so on. What new patterns emerge?

- Query the data and create new themes from selected sets of crimes. For example, if you have theft data with a time-of-day attribute, you could create a theme for car thefts that occurred between midnight and 3:00 A.M.

ANALYZE GEOGRAPHIC DATA

Below are steps that were taken by the student group in the case study. Use this list as a guide as you analyze your data.

Where is crime occurring?

- Look for patterns with types of crime, locations, and times of day or year.

- Look for relationships of crime to other geographic features such as arterial streets or freeways; proximity to certain types of businesses, schools, or other public places; areas of low or high income; and so on.

- Create a density grid showing the overall concentration of crimes.

When is the best time for the robbery task force patrols to take place?

- Create three different maps, one for each watch.

- Compare the maps, looking for similarities and differences in the distribution of crime in each watch. Hypothesize reasons for your observations.

- Compare the maps to determine the greatest areas of need at different times.

What specific areas need to be targeted with task force patrols?

- Draw polygons delineating clusters of crime or high-crime areas.

- Evaluate the number and type of crimes occurring in each hot spot.

- Compare the hot-spot characteristics to determine priorities or special needs (e.g., do any hot spots occur near schools, parks, hospitals?).

Where are the task force patrols (made up of beats) in relation to the hot spots?

- Compare the location of task force patrol zones with the actual crime hot spots.

- Consult with police to find out about additional considerations (e.g., targeting nearby residential areas where the robbers may live rather than the areas where the robberies are taking place) used to determine the actual locations of the task force patrol zones.

Where have crimes occurred recently within each task force patrol zone?

- Create several different maps to see which are most helpful for officers as they work their nightly patrols.

- Create a series of large-scale maps for each task force patrol zone showing the streets and locations of recent robberies. Include a table showing key details about each crime and the corresponding report number.

How effective was the task force in stopping robbery?

- Create maps with crime data from during and after the task force patrols.

- Compare the three sets of maps and statistics (before, during, and after) to evaluate the extent that robbery rates were affected by the increased patrols.

Where are the new robbery hot spots that should be targeted next?

- Repeat the analysis process, mapping new hot spots within the most recent data.

You can conduct a simple analysis of crime data by visually looking for clusters of crime points on a map. One advantage to this technique is that it can be done without any special software extensions. A disadvantage is that geographic patterns may not be as evident as they would be if you created density grids from them using a tool such as the ArcView Spatial Analyst extension.

If you convert the crime points to rasters using ArcView Spatial Analyst as the students did in the case study, you will be able to analyze the data using a number of quantitative techniques. A raster is a data set made up of a grid of equally sized square cells with each cell assigned a value. To convert the crime point data to a raster, you need to specify a cell size and a search radius method to use for interpolating the point locations. (In the case study, students used a cell size of .25 mile and kernel density for the search radius.) ArcView Spatial Analyst interpolates the crime point locations to come up with a value for each cell that represents the density of crime in that cell.

Once you have crime rasters, you can use ArcView Spatial Analyst or Crime Analysis Application tools to analyze them. For example, you can display crime density using a continuous color ramp that highlights the low- and high-crime areas. Or, if you have the Crime Analysis Application extension, you can have the computer quickly find the hot spots based on the cell values. For example, Bishop Dunne students quickly found hot spots in their robbery rasters using a slider bar that sets a threshold. With this tool, cells with crime density values above the threshold are displayed but those with lower values are not. The "watch comparison" maps in the case study (page 22) are a good example. You can also compare multiple rasters using the map calculator to subtract the values in one raster from those in another, for example to determine change over time, or to compare the density patterns of two different crimes. This allowed them to quickly see patterns within the data as well as perform quantitative analysis.

ArcView extensions useful for crime analysis
The following extensions are available for PC users only, running ArcView 3.1–3.3.

SOFTWARE	WHAT YOU CAN DO WITH IT	WHERE TO GET IT
ArcView Spatial Analyst extension	Create density rasters to quantitatively analyze spatial patterns within data. Perform quantitative analysis with the map calculator (e.g., subtract one raster from another to find change in crime patterns over time).	*www.esri.com* or your local ESRI Business Partner. Educational pricing is available. PC only.
ArcView Crime Analysis Application extension	Use wizards to efficiently analyze crime rasters. Works in conjunction with the ArcView Spatial Analyst extension.	Free download from *www.esri.com*. Also available on free ESRI Law Enforcement CD.

ACT ON GEOGRAPHIC KNOWLEDGE

Once you have analyzed your crime data, it's time to put this valuable information to use. In the case study, for example, the Bishop Dunne Catholic School students presented their analysis of robbery patterns before, during, and after the task force patrols to the police to help them evaluate the effectiveness of the patrols. The actions you decide to pursue will depend on your community and the nature of your analysis, partnerships, and available time and resources.

Possible action steps

- Create maps for the police department illustrating recommended patrol areas so they know where and when to focus their patrols.
- Create summary maps to share with neighborhood crime-watch groups at meetings.
- Post maps of criminal hot spots in community centers and other public locations to make the community aware of the issue.
- Post information about your project and what you learned on a community Web site.
- Write an article or press release about your results for publication in the local community newsletter or newspaper.
- Work with your local law enforcement agency to help you use your GIS results to start a crime-watch group in high-crime areas.
- Use your results to educate groups of people about the geographic distribution of crime in your community.

Tips

- Develop partnerships with groups such as homeowners associations, crime-watch groups, or the local police department, and work with them to develop an appropriate action plan.

- Remember, criminal data is sensitive information and should be analyzed and used sensitively. For example, some businesses or neighborhood associations may not want crime maps posted in public view because it could hurt business or cause people to panic or overreact. Also, criminals may use publicly posted crime maps to their advantage. It is best to work with law enforcement officials to determine the best way to provide information to those who need it.

Next steps

Did your recommendations make a difference in your community? Once you have completed your selected action on your initial question, check back for additional criminal data. Create maps of the time period before, during, and after action was taken on the problem. The students of Bishop Dunne did this and found that the robbery task force did make a difference in their community. Robberies dropped in locations where the task force was in place and stayed the same or increased in other areas. A next step for the students in Dallas is to map crime data for their neighborhood for the past ten years. The Dallas Police Department hopes this will help them see trends in how the neighborhoods have changed for the better as a result of various programs.

Often, as you find the solution to your problem you also find many more questions. Are there other related crimes that have been affected as a result of your action? What else can be done with this new information? The investigation has only begun!

MODULE 2 ACKNOWLEDGMENTS

Thanks to Brad Baker and Christine L. Voigt of Bishop Dunne Catholic School in Dallas, Texas, for contributing this module's case study.

Dallas, Texas, robbery and police beat data provided courtesy of the City of Dallas Police Department.

Roads and arterials data provided courtesy of the North Central Texas Council of Governments.

A war on weeds

When a plant species from one part of the world is introduced into an ecosystem in another part of the world, unexpected things can happen. If the growing conditions in the new location are similar to those in the plant's native habitat, the plant may quickly spread, especially when natural controls such as diseases, insect pests, or animals that eat the plant are not present in the new environment.

For thousands of years humans have purposefully or inadvertently moved plants from their native environments to other places. Commonly moved plants include those used for food (wheat, potato, parsley), textiles (cotton, flax), medicines (rosemary, aloe vera), or ornamentation (rose, tulip, cactus). These are examples of humans' intelligent transplantation of nonnative plants.

Throughout western North America weeds are literally taking over whole landscapes, crowding out native plants, disrupting local habitats, ruining agriculture and rangeland, and drastically reducing the biodiversity that is essential for a healthy environment.

CASE STUDY

Mapping noxious weeds
High school students in Shelley, Idaho, took to the fields, ranges, and riverbanks with GPS units to map the locations of harmful weeds across thousands of acres in Bingham County. Find out why their maps are essential to county and state weed control officials and how their project grew to include an ongoing summer internship program.

EXERCISE

Use GIS to map a leafy spurge infestation and compute its area
In this exercise you'll add a table of GPS data to a map and use ArcView software's draw, summarize, and calculate functions to analyze the full extent of an invasive weed infestation in a farmer's fields.

ON YOUR OWN

Explore ways to develop partnerships with state and county agencies as well as with local business people and environmental organizations to gather data on invasive plant species in your area.

Mapping noxious weeds

Bingham County, Idaho

In 1999 state officials in Idaho declared a war on weeds. Noxious weeds were such a problem that a program to eradicate or control their spread had to be initiated. The state required each county to submit electronic maps showing all areas of noxious weed infestation. This was a daunting task to Bingham County Weed Superintendent Paul Muirbrook because his department did not have a computer or an accurate GPS receiver, and he was not familiar with GIS technology.

Muirbrook's best hope was effective public education. For the past four years, students in Shelley High School's Solutions class have been using initiative and GIS to map the weeds. As of summer 2001, students have produced maps of more than 86,000 acres of land and 250 miles of irrigation canals within the county. Students also are sharing what they learn with other students, the public, and officials at all levels of government.

Four students from the Shelley High School Solutions class traveled to Boise in February 2002 to meet with Idaho Governor Dirk Kempthorne who co-chairs the Federal Advisory Committee to the National Invasive Species Council. Here, they are gathered in the governor's office along with teacher Mike Winston and Bingham County Weed Superintendent Paul Muirbrook. The students hope to convince Governor Kempthorne to make student participation a key part of his national initiative for combating invasive weeds.

WHAT IS A NOXIOUS WEED?

According to Idaho state law, a noxious weed is a plant species that meets four criteria: (1) it is present in but not native to Idaho, (2) it is potentially more harmful than beneficial to the state, (3) eradication is economically and physically feasible, and (4) the potential adverse impact of the weed exceeds the cost of control. Weeds that are native to the state may be just as harmful as noxious ones, but they are referred to as invasive weeds.

Bingham County, Idaho, is located in the eastern part of the state. The county seal reflects the county's diverse rural landscape that includes agricultural sugar beet fields on the Snake River plain, rangeland and desert, waterways, mountains, and a variety of recreational areas. Noxious weeds are typically spreading along waterways and roads.

Noxious and invasive weeds cost the Idaho economy millions of dollars each year by degrading wildlife habitats, choking streams and waterways, crowding out beneficial native plants and agricultural crops, poisoning and injuring livestock and humans, and impeding the use of recreation areas. In Bingham County, approximately fifty plant species are considered noxious weeds. Twenty species, including both noxious and invasive plants, are poisonous.

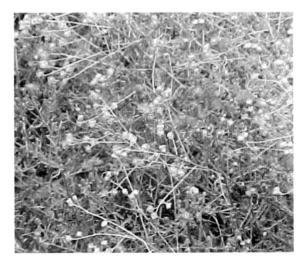

Bureau of Land Management/photo by Jerry Asher

Leafy spurge (left) and spotted knapweed (right) are two of the fifty species of noxious weeds threatening Bingham County. Students researched and learned to identify them all. They learned, for instance, that a leafy spurge plant can throw its seeds as far as 15 feet away from its stem, and that the seeds can remain viable for fifty years. People coming into contact with the plant can get a rash. Leafy spurge is native to Europe, while spotted knapweed and its cousins, diffuse knapweed and Russian knapweed, are native to Eastern Europe and Asia. Students found extensive colonies of knapweed along irrigation canals in Bingham County.

Noxious weeds typically are opportunists that spread easily by natural or inadvertent human means. For instance, weed seed can be spread across entire agricultural fields by planting crop seed contaminated with weed seed, irrigating with water contaminated with seed, or using farm equipment with weeds caught in it. Weeds also spread naturally via wind, water, and birds.

Weed seeds can be unintentionally transported by vehicles, livestock, wildlife, hikers, and horseback riders when plants or seed pods stick to tires, fur, clothing, and shoes.

Bureau of Land Management/photo by Ben Roche

Bureau of Land Management/photo by Jerry Asher

Chemical and biological controls are sometimes used to kill weeds before they reach maturity and produce new seeds. In areas like campgrounds or wildlands where native species could be harmed by herbicides, however, the only recourse is to pull weeds up by hand.

In general, noxious weeds harm wildlife and livestock by crowding out the native or cultivated plant species that normally provide food and shelter. Other effects vary depending on the species of weed. For example, black henbane, a European member of the nightshade family, can cause serious illness or death to sheep and other livestock. On the other hand, sheep and goats are sometimes used to help control leafy spurge, which these animals can safely eat.

GIS helps students find solutions

Shelley High School teacher Mike Winston first learned of GIS technology several years ago. He and a colleague were intrigued with the educational and career opportunities GIS could offer students in their Solutions class, where students are encouraged to identify and solve local community problems in areas of potential career interest. Students generally work with mentors from the community who have expertise in their area of interest. Over the course of the school year, they also develop a sense of belonging to the community, acquire marketable skills, and contribute data needed for making sound long-term decisions pertaining to natural resources and other community issues.

The first GIS project the students undertook was to map underground water levels in the Snake River Aquifer for the Idaho Department of Water Resources. The results of the project were encouraging, but the teacher who had facilitated the project left Shelley High School the next year. Mr. Winston made the decision to continue using GIS in the Solutions class even though he personally did not know how to use ArcView software. By having the experienced students teach ArcView to the new students each year, he has been able to continue supporting GIS and related technologies for student projects. He also located GIS experts in a nearby community who mentor the student teachers and help answer questions.

In 1999, when students in the Solutions class learned about the state's noxious weed mapping requirements, they realized this was an opportunity to learn technology skills while performing a valuable service to the county and state.

Getting ready

The first step was to contact the state weed department to find out about their standards for weed data. Mr. Winston knew that any weed data students collected would be worthless to the department unless it met their requirements. The state weed officials enthusiastically came to Shelley High School and taught the students the necessary information, including how to use a GPS unit, how to forecast satellite positions to determine the best times to collect data, how to incorporate a data dictionary (a scroll-down menu used to answer questions about weed infestations), and how to download data into ArcView. Local weed experts helped out by teaching students how to identify noxious weeds.

Partnerships and grants were essential to the project. Local and state weed experts taught students weed identification and GPS field data collection techniques. Equipment and software were obtained through grants and donations. Students learned GIS from other students and, on their own time, several students completed a GPS field data collection training and certification course. The certification had two objectives: to help students know what specific skills they would need for the project, and to provide a means for them to determine when they had reached that level of competency.

Community Geography: GIS in Action

The class soon realized they would need substantial financial and technical assistance from a series of partners in order to acquire the GPS equipment, GIS software, and skills needed to successfully collect the weed data. A GPS unit with the required accuracy level—within 3 feet—would cost approximately $5,000, and several units would be needed. Mr. Winston applied for grant money to fund the project, and the school ultimately received a substantial grant of $100,000 from the Idaho Department of Education. ESRI donated GIS software for the project.

Where are the weeds?

With the resources in place, students were ready to start mapping weeds. They began by systematically choosing areas to map. In the field, they recorded the exact location of weeds using field GPS units. At the end of each day of field-work, they downloaded the data into ArcView.

Once they had some data, students started creating electronic maps for the county and state weed officials. Part of their learning process was to explore the best way to present the weed data on maps. Because Superintendent Muirbrook was unsure at first what information he wanted the maps to show, students experimented with showing weed locations on land ownership maps, watershed maps, USGS 7.5-minute topographic quadrangle maps, and so on. In the end, they put all the possibilities together and allowed the superintendent to select the data layers he wanted.

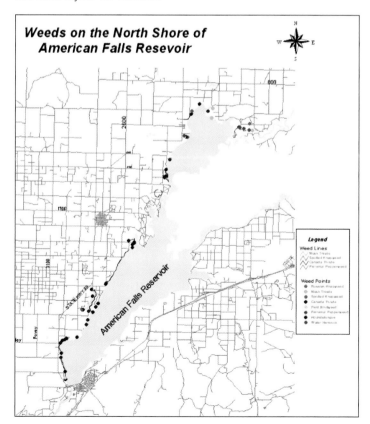

Students mapped weeds along the shore of the American Falls Reservoir. In ArcView, weed locations were displayed as points on a map. Students experimented with a number of map layouts before submitting the results to the local weed officials.

Fighting a war on weeds

The students' project was so successful the first year that Superintendent Muirbrook obtained funds to hire two students to continue during the summer. By summer 2001, eighteen students were busy mapping weeds with Bingham County and a new partner, the U.S. Department of Energy's Idaho National Engineering and Environmental Laboratory. By the end of summer 2001, students had far exceeded their original goal of mapping 10,000 acres. They had mapped 86,700 acres and 250 miles of canal and stream banks.

Altogether they documented more than 540 weed infestations, and their story was reported in the local newspapers. Weed officials and other partners were delighted to receive the valuable hard-copy maps and digital data the students provided. Local officials used the maps to understand the spread of noxious weeds in the county and to identify priority areas for spraying herbicides. State officials used the maps, together with maps submitted by other counties, to analyze the status of noxious weed infestations statewide.

The colored areas on this map represent the areas inventoried by students during 2001. Areas inventoried inside the county included the Sterling Wildlife Management Area, Morgan's Bridge and Desert Well No. 2 areas, the banks of American Falls Reservoir (the large water body in the lower left), rivers, and canals. Weeds were also mapped in sugar beet fields in the southwest part of Bingham County and adjacent Power County (represented on the map by the road network).

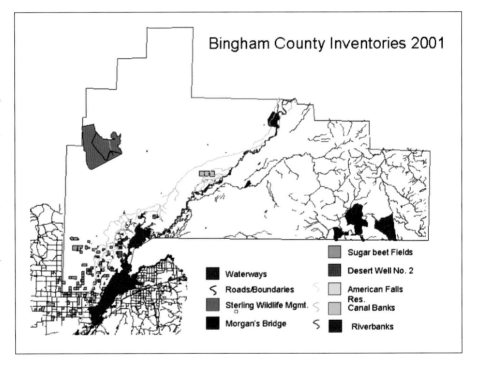

Bingham County Inventories 2001

Waterways

Roads/Boundaries

Sterling Wildlife Mgmt.

Morgan's Bridge

Sugar beet Fields

Desert Well No. 2

American Falls Res.

Canal Banks

Riverbanks

Students planned to map a significant portion of the area known as "Morgan's Bridge" (yellow). Surprisingly, they were able to complete weed mapping of a much larger area (including green).

A number of additional partners supported the students as the project progressed. These included the Blackfoot High School Solutions class, the Black Snake Cooperative Weed Management Agency, private landowners, several businesses and associations such as the East Idaho Grazing Association and the Aberdeen Springfield Canal Company, and state government agencies such as the Idaho Departments of Public Lands and Fish and Game. In this photo, a student confers with an official from the Bureau of Reclamation.

To reach the weed infestations, students used various and creative transport. Mostly they drove pickup trucks and walked, and in 2001 alone, they walked more than 250 miles along canals. Other times they traveled by four-wheel off-road vehicle, horse, or in the Sheriff's boat.

An impromptu swim was a welcome reward for a long, hot day of data collection.

Educating the public

Making maps was only the beginning for the Shelley High School students. Along the way, they have enthusiastically found ways to share their newfound expertise on mapping noxious weeds and to educate the public. In one project, students created a weed identification calendar featuring a different weed for each month. They developed a booklet containing a nontechnical weed identification key, illustrations, and descriptive information. In another project, they partnered with a local third-grade class and taught the younger children how to use GIS.

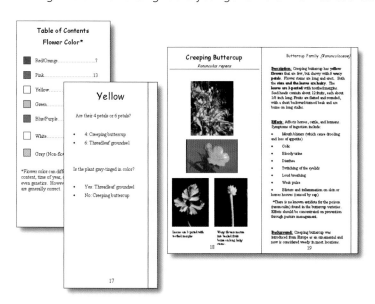
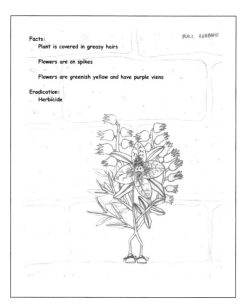

Shelley High School students developed nontechnical materials to help other young people and their parents identify poisonous weeds that could be as close as their own backyard. The plant identification booklet (left) contains a plant identification key and descriptions organized by flower color. The children's coloring book (right) includes weed-plant characters and simple descriptions.

In August 2001, a group of eight students met with representatives of the Idaho Department of Agriculture and presented a slide show that profiled each of the noxious weeds they found. A few months later, the students received a special invitation from Idaho Governor Dirk Kempthorne to attend the Idaho Weed Controllers Association annual convention in Boise. Three of the Solutions class students made the trip, accompanied by their teacher and Bingham County weed officials. The students presented their project, explaining how they used GPS to locate and map noxious weed infestations in Bingham County, and received a standing ovation.

Students returned to Boise in February 2002 to meet directly with the governor and discuss how their project might be used as a model for students around Idaho and the United States to map noxious weeds while learning important career-oriented skills. Two months later, they traveled to Washington, D.C., to present their project at the White House's Global Science and Technology Week. They were invited to participate by the Department of Commerce, which has a program supporting the use of geotechnologies like GIS and GPS in public schools.

In late April 2002, four of the students were invited by the Department of Commerce to present their weed inventory and other mapping projects as part of the White House Global Technology Week conference in Washington, D.C. One of their presentations was a keynote address at a luncheon on Capitol Hill. In this photo, the students proudly pose with U.S. Secretary of Commerce Donald Evans showing off their certificates of participation from President George W. Bush.

Next steps

Mr. Winston and his Solutions class students plan to continue mapping Bingham County's noxious weeds for years to come. For example, a five-year plan has been developed that calls for students to revisit mapped areas to evaluate the effectiveness of weed-control measures. They also expect to undertake more spin-off projects as opportunities arise. One completed spin-off project was mapping areas within sugar beet fields where beets were growing poorly. Weather patterns, a plant virus, and overspray of herbicides from adjacent properties were suggested as possible causes. Based on their research, the students concluded that the beets were harmed by noxious-weed-killing herbicidal spray drifting into the beet fields from adjacent lands. Other projects will no doubt focus on increasing public awareness of the issues surrounding noxious weeds and the benefits of GIS and GPS technologies in dealing with them.

SUMMARY

ASK A GEOGRAPHIC QUESTION	• Where are the weeds in Bingham County now? • Where are they coming from? • By what means are they spreading? • How quickly and in what areas are the weeds spreading? • Which locations are best to focus mitigation efforts such as herbicidal spraying?
ACQUIRE GEOGRAPHIC RESOURCES	• Learn standards and requirements for the data to be collected. • Form partnerships and obtain equipment, software, and other necessary support. • Complete GIS software and GPS field data collection training. • Complete training on weed identification and knowledge. • Identify which county areas are priorities for mapping. • Systematically collect weed data in the field. • Download weed locations and attributes from the GPS to ArcView.
EXPLORE GEOGRAPHIC DATA	• Explore the patterns of weed locations. • Experiment with various orientation and interpretive themes to display with the weed data, such as roads, water bodies, aerial photographs, and parcel ownership maps.
ANALYZE GEOGRAPHIC INFORMATION	• Describe the overall distribution patterns for each species of weed. • Visually identify the areas with the greatest concentration of weeds. • Analyze changes in weed distribution over time by comparing data from one year to the next.
ACT ON GEOGRAPHIC KNOWLEDGE	• Provide local and state weed officials with usable hard-copy and electronic maps. • Prepare a slide show about the project and present the methods, findings, and benefits to professionals and public officials who are concerned with noxious weeds. • Create materials to educate the public about weeds such as identification keys, lists of poisonous plants, calendars, coloring books, and so on. • Partner with the local elementary school to share information about weeds and GIS. • Attend workshops and present findings at appropriate professional meetings and conferences. • Meet with local-, state-, and national-level government officials to relate the benefits of the project and discuss ways to expand it to other schools.

Use GIS to map a leafy spurge infestation and compute its area

Exercise

Precise locations on the earth can be determined using Global Positioning System (GPS) equipment. For example, students in the case study took GPS units into the field and used them to record the locations of weeds they found along roads, streams, crop fields, and other places in the environment. After data is collected with a GPS unit, it is normally downloaded into a computer for display in a spreadsheet, database table, or GIS software such as ArcView. In this exercise, you will take a set of GPS data collected by the Shelley High School students and add it to ArcView as a table. You will use location coordinates in the table to create points on a map representing the weed infestations.

Because you are not able to actually travel to Bingham County, aerial photograph images of the study area will help you to better understand the landscape. In your analysis, you will map and measure the estimated area of infestation in acres. You will use the information from the infestation size attribute that students collected in the field.

The ✎ icon indicates questions to be answered. Write your answers on a separate sheet of paper.

ASK

A Bingham County farmer has received the following notice from the local agricultural extension agent:

"Leafy spurge is a nonnative, noxious weed that has been found in our county. It can quickly spread over large areas, crowding out native and agricultural crop plants. It is toxic when eaten by most animals (except sheep and goats). Fish and other aquatic life can be poisoned when the toxic oil produced by the plants makes its way into streams. An established leafy spurge infestation is difficult to eliminate because roots may grow 30 feet deep and seeds spread easily. A mature leafy spurge plant can produce fifty seed capsules with three seeds per capsule. When the capsules dry, they explode and shoot seeds up to 15 feet away, and then seeds may roll or bounce even farther. Seeds also can be carried by birds, insects, animals, farm equipment, harvested crops, or water."

The farmer, who grows sugar beets, believes he has identified leafy spurge *(Euphorbia esula)* in his fields. He has asked you to map the weed infestation and collect data about it. He wants to know the area of the infestation in acres and the proportion of the infestation to the total area of his fields.

EXPLORE

PART 1   EXPLORE BINGHAM COUNTY DATA AND MAP NEW GPS DATA

Before you go out to the farmer's property with your GPS unit, you decide to review your basemap of Bingham County and the weed data already collected.

1   Start ArcView. Click the File menu and choose Open Project. Navigate to the exercise data folder (C:\esri\comgeo\module3) and open *weeds.apr*.

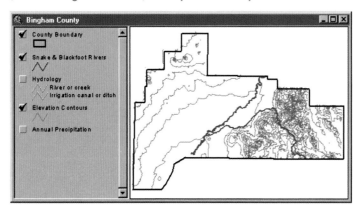

The view contains a basemap of Bingham County that shows elevation contours and the two main rivers flowing through the region—the Snake River and the Blackfoot River.

2   Answer the following questions about the map. Use the Zoom In and Identify tools if necessary.

2a   Compare and contrast the geographic relief in the eastern and western parts of Bingham County. How does the map support your reasoning?

2b   Which is the Snake River and which is the Blackfoot?

3   If necessary, zoom out to display the entire county again. Turn on the Hydrology theme.

3a   Describe three distinct regions of Bingham County based on the distribution of the hydrologic features.

3b   Rank the three regions according to where you would expect to find the most agricultural activity. Explain your reasoning.

4   Turn on the Annual Precipitation theme and make it active. From the Theme menu, choose Hide/Show Legend.

4a   Does the annual precipitation map support the regions you described in question 3a and the ranks you gave them in question 3b? Explain.

Your database includes road, landowner, and previously collected leafy spurge data for Bingham County. You will add these themes to the Bingham County view.

5   Click the Add Theme button. Navigate to the exercise data folder *(C:\esri\comgeo\module3)* and add the three themes listed in the table below. Use Theme Properties to change each new theme name.

THEME	THEME NAME
bing_ls.shp	Leafy Spurge
landowners.shp	Landowners
roads.shp	Roads

Before continuing with your data exploration, you will take a few moments to organize and symbolize the new themes. If you are using ArcView 3.0a, you will need to load the legends for the new themes. If you're using ArcView 3.1 or higher, they were loaded automatically, so skip to step 7.

6   For each of the three themes, open the Legend Editor and click the Load button. Load the appropriate legend file from the exercise data folder. For example, for the Roads theme, load *roads.avl.* For the landowners theme, accept the default settings in the small Load Legend window. Remember to apply the changes each time. Close the Legend Editor when you are finished.

7   Reorder the new themes so that your table of contents matches the one shown below.

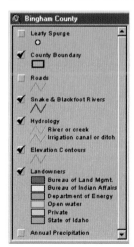

8   **Turn off the Hydrology and Elevation Contours themes. Turn on the Landowners theme.**

8a   Describe the general pattern of public and private landownership in Bingham County.

8b   Name the three landowners who control most of the publicly owned land in Bingham County.

9 Turn on the Leafy Spurge theme.

✳ **NOTE:** This theme contains the data collected by the students in summer 2002, and may represent only a portion of the leafy spurge infestations existing in Bingham County.

 9a What observation can you make about the distribution of the leafy spurge points in your database relative to landownership?

When the Shelley High students collected the leafy spurge data, they observed the density of plants growing in each place. They recorded the density information in an attribute named Coverage. In the next steps, you will classify the leafy spurge points by this attribute.

10 Open the Legend Editor for the Leafy Spurge theme, and make the following changes:

• Choose Graduated Color for the Legend Type.

• Choose Coverage for the Classification Field.

• Hold down the Shift key and select all four symbols in the Legend Editor.

• Hold down the Shift key and double-click the symbols to open the Palette. In the Marker Palette, change the symbol to an outlined circle and change the size to **12**.

• In the Legend Editor, replace the numeric labels with the words Sparse, Light, Medium, and Heavy, as shown in the picture below.

• Change the color ramp to Purple monochromatic.

11 Make sure your Legend Editor looks like the one in the following picture, then click Apply.

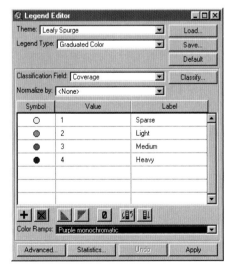

You will save this legend so that later, when you have the data from the farmer's field, you can quickly display it using the same legend.

12 Click the Save button. Navigate to the exercise data folder (or another location specified by your instructor) and name the file ls_cover.avl. Click OK. Close the Legend Editor and Symbol Palette windows.

13 Click the Zoom to Active Theme button. Visually analyze the leafy spurge points, turning themes on or off and zooming or panning the map as needed.

13a What geographic patterns do you see, if any, in the coverage density of leafy spurge?

13b What observations can you make about the location of the leafy spurge points relative to features other than landownership?

14 Close the Bingham County view. Save the project. If you are working in a classroom environment, name the project weeds_abc.shp where "abc" represents your initials, and ask your instructor where you should save the project.

Now it is time to familiarize yourself with the location of the farm you will be visiting.

15 Open the view named Farm 1.

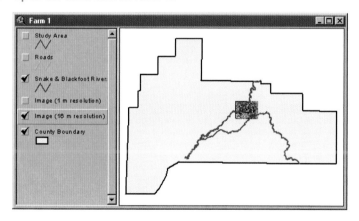

You see the Bingham County boundary, Snake and Blackfoot Rivers, and an image of a small portion of the county.

16 Make sure the Image (16 m resolution) theme is active, and click the Zoom to Active Theme button.

The features in the image vary in size, shape, and shade of gray. The circles you see are agricultural fields created by irrigation systems that revolve around a center point. Other fields are rectangular or irregular in shape. The straight, narrow lines of roads, and the meandering line of the Snake River, are also visible in the image.

16a Make three observations about the landscape you see in the image.

17 Turn on the Roads and Study Area themes. The study area marks the location of the farm you will be visiting.

18 Zoom in to the study area.

When you zoom in to the extent of the study area, you see a grid of cells instead of a clear picture. In this image, each cell represents approximately 16 square meters of area on the earth's surface. To see more detail, you will display an image with a much smaller cell size of 1 square meter.

19 Turn on the Image (1 m resolution) theme.

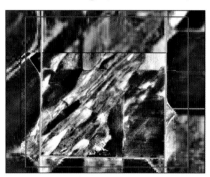

20 Use the zoom and pan tools to explore the image.

20a List three features you see on the 1-meter-resolution aerial photograph.

21 When you are finished exploring the image, zoom out to the extent of the study area.

ACQUIRE

Next, imagine that you have traveled out to the farmer's property with your GPS unit. Instead of using latitude and longitude to describe the locations on the earth, you set up your GPS to use a pair of coordinates called northing and easting. These coordinates represent a location in the universal transverse Mercator (UTM) coordinate system, which is an alternative to latitude and longitude.

As you systematically walk through the crop fields, you stop and take a reading whenever you come across leafy spurge plants. At each location, you record their location on the earth's surface, the estimated size of the area covered by the plants, the density of the plants, and some other attributes.

Back at your desk, you download the data from the GPS unit into your computer. Now you are ready to bring the new data into ArcView.

22 Make the Project Window active and click the Tables icon. Click the Add button, navigate to the exercise data folder *(C:\esri\comgeo\module3)*, and add the table *farmers_field.dbf.*

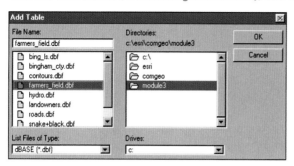

23 Scroll through the table and answer the following questions.

23a How many records are in the table?

23b What type of GPS receiver was used to collect the data?

23c Which two fields contain information about the location of the weeds?

With ArcView, you can create points on a map from the northing and easting coordinates. To do this, you will add a special kind of theme to your view, called an Event theme.

24 Close the farmers_field.dbf table.

25 Make the Farm 1 view active. Choose Add Event Theme from the View menu. In the Add Event Theme window, make sure farmers_field.dbf is selected in the Table box. For the X field, choose Easting. For the Y field, choose Northing. Make sure your dialog looks like the picture below and then click OK.

A new point theme is added to the view table of contents.

26 Turn on the Farmers_field.dbf theme. Change the name of the theme to **Leafy Spurge Coverage**. Open the Legend Editor for the theme and load the legend you saved earlier (*ls_cover.avl*). Apply the changes and close the Legend Editor.

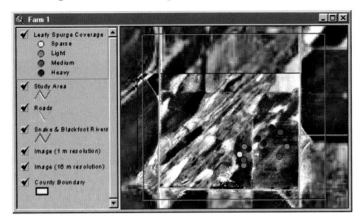

 26a What patterns do you see, if any, in the leafy spurge coverage data?

27 Click the Zoom to Active Theme button to zoom in on the infestation. You see the light-colored tracks of a farm road running north–south between two fields. The rows of crops form a pattern of regular stripes across the fields.

One of the things the farmer has asked you to determine is the total area of the leafy spurge infestation in his fields. Next you will map the leafy spurge points according to the estimated acreage of each group of plants.

28 From the Edit menu, choose Copy Themes, and then choose Paste. Change the name of the duplicate Leafy Spurge Coverage theme to **Leafy Spurge Size**.

29 Open the Legend Editor for the Leafy Spurge Size theme, and make the following changes:

• Choose Unique Value for the Legend Type.

• Choose Sizeinfest for the Values Field.

• Change each symbol to an outlined circle with a size of **12**.

• Change the colors to mustard, gold, red, and blue to match the graphic below.

• Change the labels to match the graphic below.

Symbol	Value	Label
○	.1	< 0.1 acre
○	1	0.1 to 1 acre
●	5	1 to 5 acres
●	o	other

30 Apply the changes and close the Legend Editor and Symbol Palette windows.

30a What patterns do you see, if any, in the infestation size data?

31 From the File menu, choose Save Project. Continue to part 2 now, or exit ArcView if you plan to continue the exercise later.

PART 2   MAP AND COMPUTE THE AREA OF LEAFY SPURGE INFESTATION

1 If necessary, start ArcView and open the project you saved in part 1. Make sure the Farm 1 view is the active window.

ANALYZE

The Leafy Spurge Size points tell you the estimated area of each infestation by their color, but their location represents only the approximate center of the infestation. You need to use polygons instead of points in order to represent the extent of each area. In the next steps, you will draw a circle around each point to represent the infestation areas. You will then use these circles to calculate the total acreage of leafy spurge in the farmer's fields.

Because each attribute value actually represents a range of infestation sizes, you will use the average of the range. The following chart shows the values from the attribute table and the corresponding area and radius of each circle, measured in feet as calculated using the formula area $= \pi r^2$.

Infestation size chart

INFESTATION SIZE IN ATTRIBUTE TABLE	MEANING OF ATTRIBUTE	AVERAGE ACRES	RADIUS OF CIRCLE WITH AREA EQUAL TO AVERAGE ACRES
.1	< 0.1 acre	0.05 acres	26 feet
1	0.1 to 1 acre	0.55 acres	87 feet
5	1 to 5 acres	3 acres	204 feet

2 From the View menu, choose New Theme, and then choose Polygon in the New Theme dialog and click OK. Navigate to the folder where you are saving your data. Name the file **year1_abc.shp** where "abc" represents your initials, and click OK.

A new theme is added to the view. The dotted line around the theme's check box means that the theme can be edited.

3 Move Year1_abc.shp below the Leafy Spurge Size theme in the table of contents so the circles you will draw will not cover up the dots.

You will first draw circles for the mustard dots.

4 Zoom in on the mustard and gold dots on the western side of the field.

5 Click and hold the Draw tool to display the menu of drawing tools. Select the Draw Circle tool.

6 Compare the Leafy Spurge Size legend and the table above to see how large a circle you need for a mustard dot. Notice you will draw a circle 0.05 acre in area, using a radius of 26 feet.

7 Place the crosshair of the Draw Circle tool on one of the mustard dots. Click and drag the mouse, watching the radius displayed at the bottom of the window. When it measures approximately 26 feet, release the mouse button.

(If you make a mistake, press the Delete key and draw the circle again. If the dot does not seem to be in the center, click the Pointer tool, then click and drag the circle to center it better.)

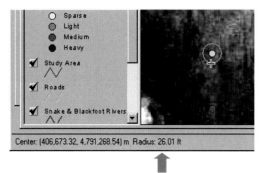

To ensure that the circle for the other mustard dot is exactly the same size as the one you just drew, you will copy and paste it.

 8 Click the Pointer tool. If your circle is not already selected (with handles like in the picture below), click the circle.

9 From the Edit menu, choose Copy Features. Then, from the Edit menu, choose Paste. You can't see the new circle because it is pasted directly on top of the first one. Click and drag the circle into position on the other mustard dot.

10 From the Theme menu, choose Save Edits.

11 Repeat the procedure in steps 6–10 to draw circles of the correct size for the gold and red leafy spurge dots. Here are some hints:

- Refer to the Infestation Size Chart for the correct radius of each circle.

- It is okay if some circles overlap others.

- Remember to zoom or pan the map to make sure you see all of the leafy spurge points.

- To copy and paste multiple circles of one size, you need only choose Copy Graphics once, after which you can simply choose Paste each time you need a new circle.

- Do not draw a circle for the blue dot, because it does not have an area associated with it.

12 When you are finished creating circles, choose Save Edits from the Theme menu.

The next step in your analysis is to determine the total area of the leafy spurge infestation in the farmer's fields. You could compute the area of each circle and then add those numbers, but any areas of overlap would be counted twice. To eliminate the overlap before computing the area, you will merge the circles into one shape.

When ArcView merges shapes, it uses an attribute field that you specify to determine which features should be combined into a single new feature. Shapes with the same attribute value are merged. Because you want all the circles to be merged, you will create a new field and assign a value of "1" to every circle.

13 Make sure that Year1_abc.shp is still active and ready for editing. Clear any selected features. Click the Open Theme Table button.

14 From the Edit menu, choose Add Field. In the Field Definition dialog, type **merge_num** in the Name box. Choose Number for the Type, and change Width to **1**. Make sure your dialog matches the picture below, then click OK.

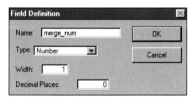

15 Click the Calculate button. In the Field Calculator dialog, type **1** in the box to complete the statement [merge_num] = 1.

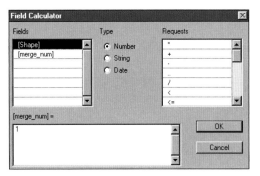

16 Click OK. All the records are assigned a merge_num attribute value of 1. From the Table menu, choose Stop Editing and save your edits (but do not close the table).

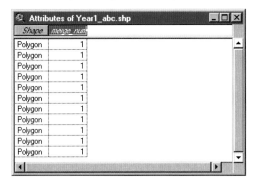

Now you are ready to merge the shapes.

17 Click the Summarize button. In the Summary Table Definition dialog, click Save As and navigate to the folder where you are saving your work. Name the file **merge_abc.dbf** where "abc" represents your initials. Make sure Field is set to Shape, and Summarize by is set to Merge. Click the Add button. Merge_Shape is added to the list on the right. Click OK.

18 Choose Farm 1 in the Summarize dialog and click OK.

19 Turn on the Merge_abc.shp theme and open its attribute table. Notice there is only one record in the table, because there is only one shape.

You will add a field to the table to store the area.

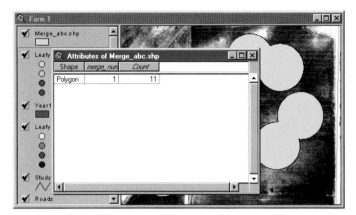

20 From the Table menu, choose Start Editing. From the Edit menu, choose Add Field, and add a numeric field named **Area** with a width of **16** and **2** decimal places.

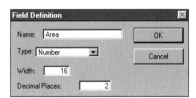

Next, you will calculate the area of the leafy spurge infestation represented by the merged circles.

21 Click the Calculate button. Type **[Shape].ReturnArea** in the Field Calculator box. Make sure your window looks exactly like the one below, then click OK.

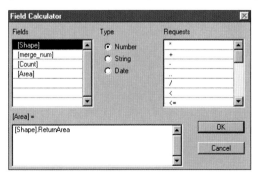

The area, expressed in square meters, appears in the table. (The units are square meters because the map units are meters. The map units are set in the View properties.)

 21a What is the area of the leafy spurge infestation?

21b Convert the area to acres (1 acre = 4,047 square meters).

21c Write two reasons why the total area figure you came up with represents an estimate, and not the actual area.

22 **From the Table menu, choose Stop Editing, and save your edits. Close the table.**

Finally, the farmer asked you to tell him what percentage of his field is infested with the weeds. To find out, you will add a theme that represents the field. It was digitized from the aerial photograph.

23 **Add the shapefile 2_fields.shp and turn on the theme. Make it active and click Zoom to Active Theme.**

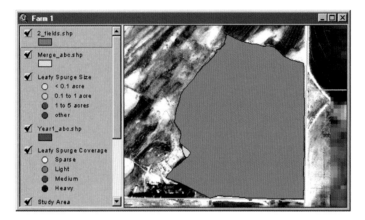

24 **Change the theme name to Farmer's Fields. Change the symbol from a solid to a transparent fill with a size 2 outline. Make the outline bright green.**

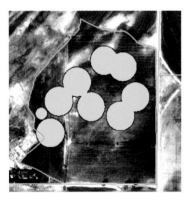

25 Calculate the area of the farmer's fields using the procedure from steps 19–21.

25a What is the area of the farmer's fields?

25b What is the area of the fields in acres?

25c What percentage of the field area is infested with leafy spurge?

26 From the Table menu, choose Stop Editing and save your edits. Close the table.

27 From the file menu, choose Save Project to save your work.

ACT

28 Write a brief report to the farmer that answers his questions outlined in the scenario at the beginning of the exercise. Include your analysis results from part 2, steps 21 and 25.

29 If you would like to print out a map to include with your report, create and print a layout now. Remember to save your work. When you are finished, exit ArcView.

The information you gave the farmer will help him develop a plan for removing the leafy spurge from his property. Congratulations on your efforts!

SUMMARY

In this exercise, you:
- Explored data to learn about the Bingham County landscape
- Created points from a table of GPS data and thematically mapped those points
- Created a new polygon theme and digitized circles representing specific area measurements
- Merged the polygons into one shape to eliminate overlap
- Calculated the area of the leafy spurge infestation and its size in relation to the whole field

ASK A GEOGRAPHIC QUESTION

Nonnative weed infestation has become an enormous problem of national importance. The survival of healthy ecosystems depends on a diversity of species. When one species destroys, dominates, or crowds out another, the delicate ecological balance is disrupted and native species are threatened. It is essential that we each take time to learn about our environment, develop an awareness of the possible environmental consequences of our actions, and decide how we might help the native species in our own communities to survive.

The extent of nonnative invasive plants and other nonnative species (e.g., nutria in Texas, or zebra mussels in the Great Lakes) is increasing at a rapid rate across the United States, according to the U.S. Bureau of Land Management and the U.S. Geological Survey. Government agencies and nongovernment groups are being organized specifically to deal with this growing problem, and most likely they would welcome an offer of technical assistance from your group. When developing your geographic question, be sure to find out which agencies or groups have ongoing projects in your area and contact them to see what kind of help they could use. Form partnerships and work with them to ensure any data you collect will be in a form that is useful to them and the community at large.

Developing your geographic question
Use the following list to guide you in developing your essential geographic question. The students at Shelley High School followed similar steps.

- Research local news sources to find out what environmental issues exist in your community. You might find announcements of funds being awarded to local agencies or new jobs being created to address the most serious or emerging issues.

- In Idaho, state regulations require county weed officers to submit digital maps of noxious weed infestations. Find out whether your state has similar requirements, and learn what people, tools, and processes are being used to gather the necessary data. The students in the case study learned that their local officials did not have the necessary tools for creating the required digital maps, and saw an opportunity to share their GIS expertise.

- The federal government and many states have laws and regulations pertaining to control of noxious weeds. Conduct research on the Internet or in the library to learn what laws or regulations your state has in place. Find out what agencies are involved with enforcing or supporting these laws, such as the state department of agriculture, the university extension service, and so on.

- Noxious weeds can have wide-ranging effects. Once you determine what harmful weeds are in your area and the kinds of environments they grow in, think about whom the weeds are affecting. The students in Shelley, Idaho, found a number of groups were affected, such as irrigation canal companies, landowner and livestock associations, private landowners, the U.S. Bureau of Land Management, and so on. Find out what kinds of maps and information about weeds these groups need.

- Conduct research on the natural association between weeds and agriculture or other forms of human activity. Consider how the information you learn could be reflected in geographic relationships between features on a map.

Your geographic question or questions will be guided naturally by the characteristics of the region you live in and the environmental impacts that noxious weeds are having on your area. For example, in Shelley, Idaho, the focus is on noxious weeds that thrive and spread as a result of agricultural activities such as cultivation and irrigation canals. In other rural areas, the issue might be noxious weeds choking lakes and ponds, transported by migrating waterfowl. In other areas, proliferation of nonnative grasses and lack of appropriate food for wildlife might be the concern.

Listed below are some possible effects of noxious weeds in your area. Use the list as a guide and think of geographic questions related to one or more of these effects.

- Taking over critical wildlife habitat areas, destroying shelter and forage.

- Disrupting migratory-bird flight paths and nesting habitats.

- Taking over large areas, resulting in reduced biodiversity in the region. Native plants cannot compete and are becoming threatened, endangered, or extinct.

- Reducing the quality of agricultural products while raising the cost of producing them.

- Causing economic impacts because government agencies at all levels as well as businesses and private landowners must increase spending to combat invasive species.

- Causing illness in domestic animals whose digestive systems cannot assimilate weeds, some of which are poisonous.

- Causing increased soil erosion because the weeds may not stabilize the soil as well as native plants do. Increased erosion may result in more sediment in creeks and streams, harming fish, macroinvertebrate habitat, and water quality.

- Reducing opportunities for hunting, fishing, camping, and other recreational activities.

- Increasing the risk of wildfire because nonnative plants may not be as wildfire resistant as native ones.

Form community partnerships

When undertaking projects dealing with a community issue, try to forge relationships with community partners. Such relationships can benefit both you or your group and the partner organizations. It is usually best to contact such organizations at the outset of your project as they may have intimate knowledge of the topic, local area, or information needed, or suggestions on how to approach the problem. Such insights could help you decide on your essential geographic question.

- Private agricultural and ranching organizations have in-depth knowledge of the private lands in your area that are used for farming or grazing. They may be able to benefit from your GIS and technical skills while sharing what they know about the surrounding environment.

Many government agencies have an interest in the millions of acres of public lands in the United States that are affected by invasive plants. Most of these agencies have programs in place that deal with noxious weeds. They may be able to connect your group with local professionals who can help with needs such as technical training and skills development. In return, your group could perform the important task of collecting weed location data, allowing the agencies to concentrate on weed control. These agencies include:

- State agencies such as departments of public lands, environmental quality, natural resources, fish and game, and agriculture

- Federal agencies such as Department of the Interior, Bureau of Land Management, Bureau of Reclamation, Department of Commerce, U.S. Geological Survey, and Natural Resources Conservation Service

- State governor's office

- County and state weed officials

Don't overlook your state or county officials who work in agricultural, weed, or pest-management departments. Many states and counties are establishing weed agencies where there may be one or two employees whose job it is to locate weeds and strategize how to control them within their geographic area. Often, these people have a huge job to do with limited resources. If you have the necessary expertise in gathering data with GPS units and in mapping it with GIS, they may be willing to provide you with whatever resources they can, because in turn you are providing personnel to help them accomplish their job.

ACQUIRE GEOGRAPHIC RESOURCES

You will need to acquire two types of data for your weed-mapping project: basemap data and weed location data. Basemap data comprises the layers that provide orientation and general information about the study area (e.g., roads, rivers, county and city boundaries). Weed location data is the location coordinates and attribute information that you will collect in the field using a GPS unit.

Basemap data

Roads, hydrology, irrigation ditches, landownership, and elevation contours are examples of basemap layers used by the Shelley High School students. You may

---

Other possible projects involving plant or animal distribution mapping

You can use geocoding skills to map and understand:

➤ Vegetation monitoring (e.g., monitor the progress of native species in revegetation or restoration projects)

➤ Vegetation succession studies and issues (e.g., agricultural land back to forest, desert, prairie; clear-cut or logged areas back to forest; and so on)

➤ Habitat mapping (e.g., "edge" habitat, which tends to have higher diversity such as good songbird habitat)

➤ Mapping effects of some other factor on plant distribution (e.g., noxious or nonnative animal or insect introduced into an area and eating or spreading a particular plant species in new ways)

➤ Bird sightings (e.g., songbird, raptor, waterfowl migration counts or nests)

---

want to include other layers that are of particular importance for your area. For example, if you suspect that hikers are inadvertently spreading noxious weed seeds, you will want to include hiking trails on your basemap. Most of the data you will want to include is readily available from a variety of sources (see the table below).

Keep the following purposes for basemap data in mind as you decide what data to acquire:

• At the beginning of the project, your basemap will help you to become familiar with the geography of your study area. This is when layers like your county boundary, roads, and elevation are helpful.

• You can use the basemap to help plan your field data collection activities. For example, you might map and identify particular parcels of land that you will inventory, or you could divide your county into a grid that you will systematically survey. Here, it's useful to map roads, trails, and other means of access.

• Basemap layers will be the backdrop on which you will map the weed location data you collect. Aerial or satellite imagery or land-use data might be useful basemap layers for this purpose.

The following table lists some common sources for basemap data:

TYPE OF BASEMAP DATA	WHERE YOU CAN LOCATE THE DATA
Basemap layers including county boundaries, streets, watershed, land owners, rivers, and so on	• Contact your town GIS planner • Visit your state's GIS Web site • ESRI Data & Maps Media Kit • Contact any GIS user in your town (utility companies, phone company, etc.) • Data sites on the Internet (see Community Geography Web site)
Topographic data	• USGS—*www.usgs.gov*—search on elevation • ESRI ArcData℠ (National Elevation Dataset)—*www.esri.com/data* • The Federal Geographic Data Committee Clearinghouse for spatial data—*www.fgdc.gov/clearinghouse* • Bureau of Land Management—*www.blm.gov*
Digital orthophoto quad (satellite image or aerial photo)	• Depending on ownership and copyright issues, you may have access to the satellite images at no charge or you may have to pay a fee to acquire it • Two excellent sources for DOQs online are *www.usgs.gov* and *www.terraserver.microsoft.com*

Places where you can look for local basemap data:

• Your town's planning office—especially if they use GIS.

• If your own town does not use GIS, your county or state government's office of economic development or regional/statewide planning may have appropriate data.

• GIS users in your town or region (utility companies, phone company, and so on) may be willing to provide some of the data you need.

• Your state's Web site may identify sources of local data that are available to you.

Data collected in the field

Collecting data in the field requires both time and skill. To ensure that the time you invest collecting data ultimately produces data that is useful, work closely with your community partners to make a plan and develop skills. The partners in the case study provided training in using GPS units, collecting attributes, determining which codes or units of measure to use for each attribute, and identifying various weed species. Eventually, a student certification program was developed to help students gain the requisite skills in GPS field data collection, map reading and orientation, and plant identification. Partners also helped students determine priorities for the locations to be inventoried.

When you go into the field to collect data on weed infestations, you will likely want to acquire descriptive information about the infestation as well as the location. The following table lists some of the attribute data collected by the Shelley High School students. When you plan your data collection, consider creating a similar table. It can be a useful reference once you start collecting data, and it can later become part of the metadata you create for your database.

FIELD NAME	DEFINITION	SAMPLE VALUE	MEANING
GPS Date	Date the weed was found and recorded with the GPS.	20020611	June 6, 2002
Site_ID	Unique identification code for each site. In the case study, these were assigned by the county weed department.	06B000073	Refers to a specific location
Species	Code for the plant species.	EUES	*Euphorbia esula,* or leafy spurge
Phenology	Code for the plant's growth cycle status when found.	FL	Flowering
Sizeinfest	Rough estimate of infestation size. For example, in the case study, students visualized a typical classroom as 0.1 acre. The estimate tells weed officials how much herbicide to bring to a site.	1	1 to 5 acres
Percent_co	Rough estimate of the density of the plants in the infestation area expressed as a percentage of the ground that is covered by weeds.	5%–25%	5% to 25% of the area is covered by weeds
Northing	Coordinates for the discreet location on the earth where the weeds were found. Northing and easting coordinates are used with the UTM zones, but locations may also be recorded as latitude and longitude.	4775785.11	Distance north of the origin in the UTM coordinate system
Easting		379000.077	Distance east of the origin in the UTM coordinate system

Tips on field data collection

The initial data you collect is the most important because it defines the database that will be used for collecting future data in the community. It also sets the protocols and standards for future data collection.

- Determine the content and characteristics of your database by talking to weed officials and others who will use your data. For example, identify the attributes (date, species name, location, infestation size, growth stage, and so on), the level of precision needed, and whether to use a number or a description.

- Prepare a data collection protocol so that everyone involved in collecting data is following the same procedures, using identical equipment, and using the same criteria for evaluating or measuring the weed infestations.

  - For example, if a rough estimate of infestation size is needed, decide on some rules of thumb (e.g., is it larger than a classroom?). Or, you could measure the length and width and calculate a rough area from that.

  - To measure coverage density with some precision, you might mark off a regular grid with surveyors tape (e.g., square feet, square meters) for a portion of the infestation and extrapolate the value for the entire infestation area.

  - To determine the location, you could take one GPS reading at the center of each infestation that will eventually become a point on a map. Alternatively, you could walk around the perimeter of large infestations, taking GPS points every 20 feet, for instance, and eventually create a polygon on a map. Or, you may decide to use both methods.

- Set up a data-recording sheet to match the data fields that you will create in the GIS project.

- Hold a training day for all participants before beginning the data collection. Include information on appropriate shoes and clothing, and how to take care not to inadvertently spread weeds further. Practice using the equipment.

Technical tips

You need to be aware of the map projection or coordinate system of any digital map data that you acquire. You will probably want to choose a map projection or coordinate system to use as a standard for your project and then request data in that format. Additionally, you will need to consider the system that your GPS will use to specify locations. For example, you may be able to choose UTM coordinates (northing and easting for your zone) or latitude/longitude coordinates.

Typically, government agencies have a standard map projection or coordinate system that they use. Many state or local agencies use State Plane or UTM coordinates. Some agencies use a system that is not widely used outside the agency. Data you acquire from different agencies may have been derived using different coordinate systems or map projections. If so, you will need to reconcile the data to display the layers in the same view in ArcView.

You can change the map projection on the fly using View Properties. Consult ArcView help for more information on map projections as well as the data projection utility or extension that comes with the software.

EXPLORE GEOGRAPHIC DATA

You probably will explore your GIS data at two different stages of your project. At first, you will want to explore the basemap data you have gathered. Later, you will create event themes with the GPS locations you collected and explore the weed data.

Preliminary explorations

Before beginning your actual analysis of the data, take time to explore the data and familiarize yourself with its content, characteristics, and potential spatial patterns. Suggestions for preliminary data exploration:

- Add themes to the view and experiment with symbology to find an effective way of representing the data. Reorder themes in the table of contents so basemap themes and weed themes are visible.

- After you have verified and corrected your data as necessary, look for geographical patterns in the weed data. Classify the weed points according to different attributes such as plant species, infestation size, and density. Compare the distributions of different weed species as well. Hypothesize the factors influencing these distribution patterns, such as soil moisture or agricultural activity, or other factors that could make a good (or bad) host environment for invasive plants.

- Compare the weed locations with geographical features in other themes. Notice whether there are features that could be contributing to weed dispersal, such as a nearby road or drainage ditch.

ANALYZE GEOGRAPHIC DATA

The type of analysis you perform will be derived from the geographic question of your project. The Shelley High students wanted to map the locations of weeds and identify spatial patterns. Possible steps you could follow include:

- Visually describe the overall distribution patterns for each species of weed.
  - Query the weed data to create separate themes for particular subsets of the data (e.g., different plant species).
  - Classify infestation points or features by various attributes (e.g., species, size of infestation, phenology).
  - Visually identify the areas with the greatest concentration of weeds.
- Use GIS tools to test hypotheses about weed distribution patterns or to quantitatively describe the patterns.
  - Create buffers of features such as hydrology, roads, or agricultural land-use areas and then use theme-on-theme selection to determine what proportion of weed points fall within or outside the buffers.
  - Add a DOQ in the background to support interpretation of the weed distribution patterns that may not be apparent from the vector data themes.
  - Digitize circles or buffers around the infestation points and calculate the total area of the infestations.
- Once you have data from multiple time periods, analyze changes in weed distribution over time by comparing data from one year to the next.

Module 3   A war on weeds   89   *on your own*

Drawing conclusions

The key to drawing conclusions is to refer back to your original geographic question, and then to step back and look at the data as an impartial, critical observer. You have probably invested significant time and resources by this point in the project and you may be tempted to draw conclusions quickly. Remember, a careful analysis of the data is critical to making good decisions that will lead to effective actions. Work with your community partners to draw conclusions that are relevant to others.

- If your geographic question is "Where are the weeds located?", can you draw conclusions about the kinds of places where weeds were found?
  - Is there something happening in the areas where weeds were found that could be weakening the native plants and thus encouraging the growth of noxious weeds (e.g., livestock grazing or an insect pest)?
  - Were the infestations of a particular species typically found in agricultural areas or areas with disturbed soil? Were they found in areas with well-drained soil (e.g., hillsides) or constantly wet soil (e.g., near streams)?
  - If several different species were commonly found together in small or confined areas like individual fields or along roadways, do the species have similar environmental needs such as wet or boggy soil?
- If your geographic question is "How are weeds spreading?", can you draw conclusions about the major access points that you can see for noxious weeds?
  - Who owns which field? Does one farmer's field have infestations while another farmer's doesn't? Does the same farm equipment move from field to field in that area?
  - Does any physical feature that might encourage a heavier growth of noxious weeds show up on your map? Is there a river or irrigation ditch that might be transporting the infestation to different areas?
  - Are there animals grazing that might be moving the seeds from one place to another?
  - Do you typically find the same weed species in widely scattered locations? For example, is one type of weed found in farm fields far apart from one another, with no sign of it in fields in between?

ACT ON GEOGRAPHIC KNOWLEDGE

The Shelley High students understood that all their hard work collecting and mapping weed data wouldn't really help fight the war on weeds unless they took action to share what they learned. Here are some possible actions you could take:

- Provide local and state weed officials with hard-copy and electronic maps designed for their needs. For example, the students in the case study experimented with a variety of map layouts and worked closely with local weed officials to develop hard-copy and digital maps that would best meet their needs.
- Prepare a slide show about the project and present the methods, findings, and benefits to professionals and public officials concerned with noxious weeds. You may wish to schedule an appointment with the town council or board

of supervisors to make a presentation of your findings at one of their monthly meetings.

- Create materials to educate the public about weeds such as identification keys, lists of poisonous plants, calendars, coloring books, and so on. Tell the local parks and recreation department that you would like to share these materials with them.

- Partner with the local elementary school to share information about weeds and GIS. Contact the media to document the event.

- Attend workshops and present findings at appropriate professional meetings and conferences.

- Meet with local-, state-, and national-level government officials to relate the benefits of the project and discuss possible ways to expand it to other groups or schools. Perhaps this step is the most critical because this is where systemic change can happen. Legislators are very interested in what their constituency has to say.

NEXT STEPS

The completion of a noxious weed survey is not the end of the project. Even with an aggressive abatement plan, noxious weeds will continue to grow and spread. See if the agency you worked with will help you develop plans for expanding the data collection to other lands or for going back to re-inventory lands on a periodic basis. You could analyze data collected over time to provide feedback on abatement efforts, quickly identify new infestations, or discover how plants are spreading. The Shelley High students developed a five-year plan with their partners so that student projects would continue to provide weed officials with updated data over time.

Any weed-mapping project you undertake will be a step forward in the daunting task of limiting the damage caused by these plants. Through such a project, you will gain valuable knowledge that you can share with your community in a variety of ways. Whether your group is large or small, you can continue studying other areas in the community, and you can connect with a larger group that has ongoing projects to address noxious weeds, such as an environmental organization or local special-interest environmental group. Remember, you have the ability to effect change!

MODULE 3 ACKNOWLEDGMENTS

Thanks to the Shelley High School Solutions class and teacher Mike Winston for contributing this case study.

Thanks to Bingham County, Idaho, for contributing the county boundary, roads, contours, hydrology, landowners, and Snake and Black Rivers data.

Thanks to the USGS for the aerial photograph images downloaded from the TerraServer Web site.

Average annual precipitation data provided by the U.S. Department of Agriculture Natural Resources Conservation Service and the Spatial Climate Analysis Service.

Hydrology data is U.S. Census TIGER/Line data.

Tracking water quality

Water pollution remains one of the most visible and persistent signs of the human impact on the natural world. Over the last twenty years, the quality of river, lake, and other surface water has improved in most developed countries, yet problems from fertilizer and pesticide contamination continue. Life depends on clean water and studying local waterways is one way we can hope to understand and preserve the quality of this crucial resource.

CASE STUDY

Monitoring seasonal changes on the Turtle River

High school students in Grand Forks, North Dakota, set out on a year-long adventure to explore the health of the nearby Turtle River. As the year of water-quality testing progressed, they gained the interest of local park rangers, who wanted to identify areas of the river suitable for fish habitat improvement. As the students expanded their research to aid local park rangers, they also made numerous educational presentations to the public about their results.

EXERCISE

Analyze Turtle River data to identify locations for fish habitat restoration

You will use GIS to thematically map tabular water-quality data about a local river, and then compare the data to a set of water-quality guidelines to determine which location on the river is most suitable for fish habitat restoration.

ON YOUR OWN

Explore the tasks and resources needed to set up a long-range study of a local waterway and use the power of GIS to store and analyze data. By outlining strategies and suggestions for data collection, data analysis, and the presentation of conclusions, this section provides guidelines for those who want to undertake a comparable long-range project.

Monitoring seasonal changes on the Turtle River

Grand Forks, North Dakota

For Roger Palmer's students at Red River High School, putting on waders and walking into icy water to conduct a water-quality test is routine. For four years, Mr. Palmer's students have been learning science through field investigations of their surroundings in northern North Dakota. Their yearlong study of the Turtle River was no exception. It began as an interesting topic for a one-week summer field-science class—"How healthy is the Turtle River?"—and evolved into an ongoing project that has attracted the attention of local environmental organizations.

The summer field-science class collected water-quality data at five sites along the Turtle River and presented their results to Turtle River State Park campers and officials. The project was such a success that Mr. Palmer decided to extend the study to involve other classes and teachers throughout the year. Biology and geology students went to the river during different weeks in the spring to collect water so that students could obtain and analyze a representative cross-section of spring runoff. In addition, a science club volunteered its time after school and on weekends to help monitor the area more regularly. By spring 2002, the students had expanded their study to a seasonal analysis of the river's water quality. Along with the first summer's data, they collected detailed data sets for the fall, winter, and spring.

Taking water samples throughout the seasons presented students with different challenges and experiences. In the spring-time, hip waders were as valuable for getting through the soft mud-filled ditches as they were in summer for keeping dry while collecting water samples.

In winter, students braved wind and freezing temperatures to get water samples. Using a gas-powered ice auger, they bore a hole through 3 feet of ice to the cold liquid below.

The students collected water-quality data from this site, known as site 5, and four others to determine the health of the river. The rocks at this site were placed here years ago to prevent river erosion.

The 48-mile Turtle River was an obvious choice for student study for several reasons. First, it is near the school. Second, the river is shallow in many areas, making it easy to cross and take water samples and GPS coordinates in the middle of the channel. Third, the river itself has five sites that are easily accessible by road and equidistant from each other, making them ideal testing sites.

Site 1 is farthest downstream, at the last bridge before the Turtle River flows into the Red River. Site 2 is just beyond where two major drainages flow into the Turtle. Site 3 is at the bottom of a ridge. Site 4 is in the Turtle River State Park, where access to the river is convenient. Finally, site 5 is at the farthest upstream bridge crossing the Turtle River.

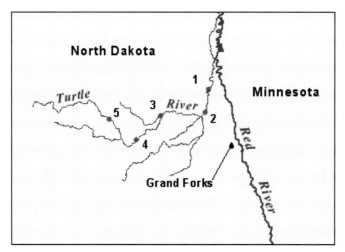

THE VALUE OF COMMUNITY PARTNERSHIPS

In the summer of 2000, as Mr. Palmer and his summer field-science class prepared for their initial study of the Turtle River, they formed partnerships with local organizations interested in their work.

First, Mr. Palmer asked researchers from the Energy and Environmental Research Center (EERC) at the University of North Dakota to instruct the students in local geology and the proper methods for collecting water samples and conducting water-quality tests in the field. Through this interaction, the students developed a list of water-quality variables that were appropriate to test.

The Turtle River State Park provided the students and teachers rustic cabin lodging for the week and a base from which to conduct their testing. As thanks for the accommodations, the students presented a multimedia program for all interested campers on the data they collected and the health of the Turtle River. From that initial positive experience, the state park became interested in the students' work.

Finally, the students needed a place to analyze the water samples they collected. That place turned out to be the Grand Forks Water Treatment Plant. Students arrived with their water samples, and after an instructional session on how to use the equipment, they went to work.

Throughout each season's testing (summer, fall, winter, and spring), Mr. Palmer's students interacted with their partners at the EERC, the Turtle River State Park, and the Grand Forks Water Treatment Plant. Although the students benefited significantly from these community partnerships, the community organizations benefited as well. The EERC was able to obtain accurate water-quality data for a local river and fulfill the educational outreach component of their funding at the same time. The Turtle River State Park benefited by having the students perform multiple interpretive programs for campers to educate them on the health of the Turtle River. In addition, the students' river data was used by the park in a trout habitat restoration project. For the generous way the Grand Forks Water Treatment Plant opened their facilities to Red River High and other students in the community, they received public recognition through various newspaper articles and news broadcasts.

Because of these partnerships, the students had access to experts in water-quality issues and high-quality analysis equipment. This allowed them to complete a much broader and informative study than originally planned for the intensive summer field-science class.

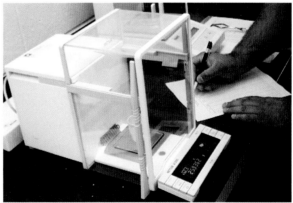

Students use analysis equipment at the Grand Forks Water Treatment Plant to conduct their water-quality tests. They recorded their results on a handwritten data table and transferred them into a computer spreadsheet so they could be mapped in ArcView.

CHEMICAL TESTING

The students identified a list of variables they were going to test. The following table is a summary of all the tests they conducted:

TEST VARIABLE	DESCRIPTION OF TEST
Alkalinity (carbonate concentration)	The measurement of how much acid a body of water can neutralize before it is adversely affected. Carbonate is an ion that forms a buffer against acids. Alkalinity diminishes agricultural productivity so that most lands that have highly alkaline soils are left as low-maintenance rangelands or wildlife preserves.
Conductivity	A measure of the dissolved salts in the water. Conductivity is determined by seeing how much electrical charge can pass through 1 milliliter of water at the tip of the probe. Highly saline water tends to promote inhabitation by only the most salt-tolerant plants and animals.
Hardness (calcium and magnesium)	Water hardness is a measure of the +2 charged ions in the water. Most commonly, these ions are calcium and magnesium and sometimes iron. Hard water leads to diminished soap activity in the home, which is the basis for all water softness sales.
Nitrogen (ammonia and nitrates)	A measure of the nitrogen in the water that is in the form of nitrates or ammonia. Nitrogen is the most common element in our atmosphere. It is released in the waste of animals and the decay of plants. It is usually the largest component in most fertilizers. High nitrate levels cause plants to become overgrown and then compete for necessary oxygen as they die and decompose.
Dissolved oxygen	The amount of oxygen dissolved in water indicates the likelihood that the water will support abundant aquatic life. Cold water and fast-moving water are generally able to dissolve more oxygen.
pH	The pH scale measures how much acid (hydrogen ion) is in the water. A low pH (0–6) indicates high levels of acid and a high pH (8–14) indicates low levels of acid (known as basic or alkaline conditions). Neutral water normally has a pH of 7.
Phosphates	A measure of the phosphates dissolved in the water. Phosphates stimulate the growth of plants and are commonly found in fertilizers. Too many phosphates cause waterways to become overgrown with plants.
Sulfates (magnesium sulfate or sodium sulfate)	Sulfates are generally nonreactive and are not considered significant in water-quality calculations. However, they can contribute to the total conductivity in water.
Temperature	Temperature measures the average kinetic energy of the water molecules. Temperature influences the location of fish significantly. Oxygen levels, growth, feeding, mating, and migration are all affected by water temperature.
Turbidity	Turbidity measures how cloudy the water looks. It is determined by submerging a black-and-white Secchi disk and measuring the maximum depth at which the white and black pattern on the disk can be seen. Turbid water does not allow sunlight to penetrate as deeply as clear water and indicates poor water quality.
Depth	Water depth is measured at the deepest part of the channel and is measured in feet or meters. In winter, northern rivers can freeze 2 to 3 feet deep. Deeper rivers contain a living space beneath the ice for all nonburrowing and non-hibernating fish, reptiles, or amphibians.
Flow	Flow is the speed of water. Using a sphignometer, it is measured at 1-meter intervals across the width of the river, at 18 inches from the river bottom. An average reading in meters per second is calculated from all the measurements taken across the river.

At site 3, students waded into the Turtle River to measure the water depth at the center of the channel.

This student measures flow using a sphignometer. A cupped wheel turns as the river current pushes it. Each turn creates an electronic click that the student counts. With the number of clicks per minute, the student can calculate the linear flow of the river at that location.

Students are mapping the bottom profile of the river by marking registration points with their GPS unit. The latitude and longitude coordinates will help them align their maps with other data in ArcView. These bottom profiles are key in determining the best location for trout habitat improvement along the river.

After the students collected their data set each season, they imported their data into ArcView and thematically mapped it. Because they used a spreadsheet program to record their data, they were able to easily join each new season's data to the existing table and map the results.

Dissolved oxygen data for the summer, fall, winter, and spring was mapped using bar chart symbology. The students observed that temporally, the level of dissolved oxygen was lowest in summer and highest in winter and spring. Spatially, they noted that site 1 had the lowest dissolved oxygen readings and site 5 had the highest.

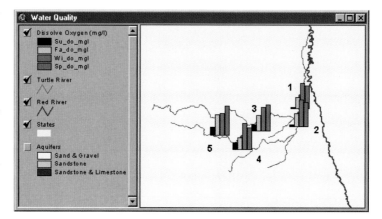

LOCAL GEOLOGY HELPS EXPLAIN THE HEALTH OF THE TURTLE RIVER

As the students analyzed their results using GIS, they referred to the geology of the area to help explain what they found. The Red River Valley, on the border of North Dakota and Minnesota, is covered with thick, rich earth that produces some of the best crops in the world. The valley lies at the southern edge of what was once a vast glacial field. Upon climatic warming, a massive lake formed at the base of the glacial field. As this lake alternately drained and stabilized, beach ridges formed along the shorelines. These beach ridges became buried under sediment and plant growth and formed many shallow sand and gravel aquifers that still exist today. One such aquifer is located between the source of the Turtle River and site 5.

The students learned that two additional aquifers affect the local region. The sandstone aquifer under sites 1 and 2 was formed from a series of dead lakes across the Dakotas and became filled with sediment during the Cretaceous period. Like the Great Salt Lake of today, many salts were caught in these closed basins, causing the aquifer to be saltier than other aquifers in the region. The sandstone and lime-stone aquifer east of sites 1 and 2 was created when the shallow oceans that existed between the Appalachian and Rocky Mountain ranges filled when the continent was young. Remains from shellfish, coral, and material from eroding mountains produced sedimentation layers that contain the large amount of limestone found in this aquifer.

The Turtle River flows over two aquifers and near a third. All three aquifers influence the water quality of the river. For example, the students found that sites 1 and 2 had the highest levels of conductivity, as a result of the increased limestone in the area.

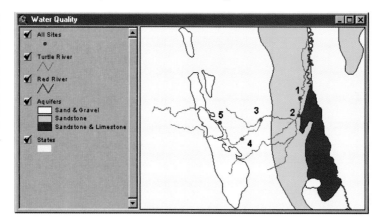

Overall, the students discovered that the water quality of the Turtle River is healthy. It corresponds to data they'd expect to see of water that is influenced by sand and gravel, sandstone, and sandstone and limestone aquifers. By recording this data, the students have a baseline that they can refer to in future years. If there are any changes to the land use around the Turtle River in years to come, students will be able to determine resulting effects on the water quality of the Turtle River.

MAKING THEIR RESEARCH KNOWN

At various times throughout the yearlong study of the Turtle River, Mr. Palmer's students gave presentations to local television stations, Turtle River State Park campers, and a variety of conferences in North Dakota and Minnesota. In their interpretive program to the state park campers, the students focused on why different areas of the river would be better for fishing than other areas.

A student presents the water-quality results, using GIS and a map of the local geology to explain the water-quality data to members of the North Dakota Academy of Science, May 2002. The students also traveled to Bozeman, Montana, to receive recognition for their research at the regional Showcase of Outstanding GIS Classroom Projects, fall 2001.

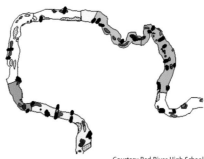

Courtesy Red River High School

The presentation on fishing caught the attention of the Red River Regional Council (RRRC), a government organization that conducts environmental projects that enhance the local economy. One of the RRRC's projects, the Red River Basin Riparian Restoration Project, shared some common goals with the students' study. Specifically, the RRRC was interested in conducting an assessment of the Turtle River to examine the stability of the river banks, location of logjams, location of in-stream fish habitats like riffles and pools, the condition of the river where it flows through the Turtle River State Park, and the overall water quality. The RRRC was interested in evaluating the student data for a possible trout habitat restoration project on the Turtle River. In the summer of 2002, Mr. Palmer delivered the water quality data to the RRRC for its analysis. At press time, the RRRC had only begun its evaluation.

The students created a bottom profile map of the Turtle River near the Turtle River State Park (site 4). Dark green areas contain deep water and the blue areas are holes in the river bottom. Logs are symbolized with a thick black line and sandbars and large rocks are brown dots and gray areas, respectively. The bottom profile map was given to the RRRC for use in determining appropriate trout habitat restoration areas. The RRRC will share this map with its partners in the restoration project: Grand Forks County Water Resource Board, North Dakota Game and Fish, and Turtle River State Park.

case study

NEXT STEPS

As the students prepare for another year of chemical testing, they also plan to expand their study to include an examination of the macroinvertebrates that live in the river. They will sample the water for the insects that make their home under rocks and logs. They plan to continue a yearly interpretive program at the Turtle River State Park and keep working with the RRRC on trout habitat restoration. Overall, the students of Red River High have laid a foundation of local research and action that can be built on for years to come.

SUMMARY

ASK A GEOGRAPHIC QUESTION	• At which sites is the Turtle River healthy?
	• Which sites will support trout habitat restoration?
	• What influences the water quality of the Turtle River?
ACQUIRE GEOGRAPHIC RESOURCES	• Obtain GPS points of all the sites.
	• Determine protocol for collecting water-quality data.
	• Collect water-quality data for each of the sites, for all four seasons.
	• Record data in a table that can be imported into ArcView.
	• Obtain basemap data of the area (river locations, state boundaries, and aquifers).
EXPLORE GEOGRAPHIC DATA	• Thematically map each water-testing variable using graduated symbols, graduated colors, and bar and pie charts where appropriate.
	• Join new data tables to existing data.
	• Visually analyze patterns in the data.
ANALYZE GEOGRAPHIC INFORMATION	• Visually analyze each test to identify temporal and spatial trends.
	• Overlay aquifer data to explain identified trends.
	• Repeat study to see how patterns change over time.
ACT ON GEOGRAPHIC KNOWLEDGE	• Summarize results.
	• Prepare and practice presentations for interpretive programs, television interviews, and local conferences.
	• Give presentations.
	• Supply Red River Regional Council with data for trout habitat restoration.
	• Write magazine articles and notify local newspapers about the project.
	• Continue water testing and expand project to include identifying macro-invertebrates.

Analyze Turtle River data to identify locations for fish habitat restoration

When collecting data at different times of the year, you must have a way to add new data to the preexisting data set. In this exercise, you will "join" data collected in the spring to the existing table of summer, fall, and winter data. You will thematically map the data and evaluate each testing site's suitability for fish habitat restoration.

The ⬚ icon indicates questions to be answered. **Write your answers on a separate sheet of paper.**

PART 1   UNDERSTANDING THE GEOLOGY OF THE TURTLE RIVER

ASK

You have been asked by a local fishing club to help them analyze water-quality data collected at five sites along the Turtle River and to provide a recommendation for the best site for a fish habitat restoration project. The fishing club gave you best-site criteria that must be met: the water must be at least 3 feet deep, the dissolved oxygen levels need to be on average greater than 10 mg/l, the pH must be as close to 7.0 as possible, and the conductivity levels should be low.

ACQUIRE

1 **Start ArcView. Click the File menu and choose Open Project. Navigate to the exercise data folder** *(C:\esri\comgeo\module4)* **and open** *water.apr.*

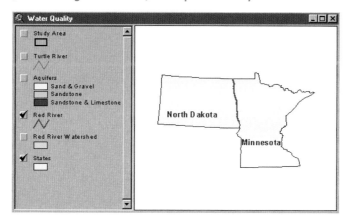

North Dakota and Minnesota are shown on the map. The Red River appears to be a boundary between the two states. First, you will explore the map to learn about the geography of the Turtle River. Then you will analyze the data collected by the fishing club.

2 Turn on the Turtle River and Red River Watershed themes. Make the Red River Watershed theme active and your view window larger.

The Red River watershed encompasses all of the area that drains into the Red River (including the Turtle River). Each major watershed is divided into smaller regions, the smallest of which is a unit.

3 Click the Zoom In tool. Draw a rectangle around the Turtle River and its confluence with the Red River.

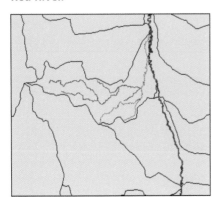

EXPLORE

4 Use the Identify tool and map to answer the following questions:

4a Name the two units in the watershed just south of the Turtle River.

4b Name the watershed unit that the Turtle River is in.

4c Which direction does the Turtle River flow?

5 Turn on the Aquifers theme and make it active. Zoom to the active theme.

An aquifer is an underground region of porous material that stores water. This data set includes three types of aquifers whose characteristics are described below:

AQUIFER	AQUIFER FORMATION	EFFECT ON WATER QUALITY
Sand & Gravel (glacial)	Formed by glacial deposits and glacial lakeshores.	Because this rinsed gravel bed doesn't dissolve into the water, it has little effect on water quality.
Sandstone	Created by metamorphic processes when sand is buried under quantities of other erosional material.	Because this aquifer has been filled with water from a dead lake (a lake with no water outlet), it has concentrated salts stored in the water. There is no effect on pH (acid/base).
Sandstone & Limestone	Formed by shallow oceans providing habitats for coral and shellfish. Eventually, these oceans were covered by sedimentation.	The calcium carbonate that makes up limestone is dissolved and changes the pH to become more basic (pH of 8 or higher).

Generally, when a river flows over an aquifer, water from that aquifer is contributed to the river. Depending on the type of aquifer, the river's chemical composition may be affected by the components dissolved in the water of the aquifer. In addition, a river can be influenced by aquifers it flows near as well as those it flows directly over.

6 Zoom in on the Turtle River.

6a What types of aquifers affect the Turtle River?

6b Use the map and the information in step 5 to describe two ways you think the aquifers may affect the water quality of the Turtle River.

When the Turtle River water samples were collected and tested, the results were entered into a spreadsheet. Now you will add that data table.

7 Click the project window title bar. Click the Tables icon and then click Add. Navigate to the module 4 exercise data and change the List of File Types to Delimited Text (txt). Add 3seasontable.txt.

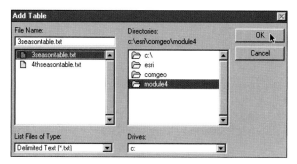

The table consists of five records and many fields. Each record corresponds to a site on the Turtle River where water-quality tests were conducted. Each field corresponds to a measurement taken at that site. Each field name consists of a season code and a variable code; this makes it easy to see which season and which variable you are examining.

CODE	SEASON
Su	Summer
Fa	Fall
Wi	Winter
Sp	Spring

FIELD NAME	VARIABLE NAME	VARIABLE DEFINITION	UNITS OF MEASURE
Depth	Depth	The depth of the river at its deepest point.	Feet
Flow	Flow	How far the water moves in one minute in the center of the stream.	Meters per second
Do_mgl	Dissolved Oxygen (DO)	The amount of oxygen dissolved in the water. Oxygen is necessary for aquatic animals to survive. Cold water, water with moderate amounts of plant life, and areas of fast-running water are high in oxygen.	Milligrams per liter (mg/l)
temp	Temperature	Measure of average kinetic energy (speed) of atomic particles. The skin senses this as warm or cold.	Centigrade
conduct	Conductivity	Amount of electricity that is able to flow through a set distance of sampled river water. It measures how many salts (ionic compounds) are dissolved in the water.	Microsiemens (μS)
ph	pH	Measure of how much acid is in the water (the hydrogen ion concentration).	pH scale (0–14)

8 Examine the attribute table and answer the following questions:

8a Which seasons are included in this data set?

8b What value and unit of measure are displayed for Dissolved Oxygen at Site #4 in the summer?

8c What value and unit of measure are displayed for Conductivity at Site #5 in the winter?

8d Based on the data, which sites have the coldest temperature and what is that temperature?

Now that you have a better understanding of the data in the table, you can map it.

9 Make the View window active. From the View menu, click Add Event Theme. If the 3seasonwater.txt file does not automatically display in the Table field, navigate to the 3seasonwater.txt table. Make sure your Add Event Theme window matches the one below and click OK.

The new point theme, 3seasontable.txt, has been added to the view.

Community Geography: GIS in Action

10 Clear any selected features. Make the 3seasontable.txt point theme active. From the Theme menu, click Convert to Shapefile. Navigate to the exercise directory *(C:\esri\comgeo\module4)* and name the new theme **3season_abc** where "abc" represents your initials. Click OK and add the shapefile to the view.

⁂ NOTE: If ArcView asks if you want to save the shapefile in projected units, click No.

11 Turn on 3Season_abc.shp and Study Area. Make Study Area active and click Zoom to Active Theme.

12 Use the Legend Editor to change the symbol color of 3season_abc.shp to red.

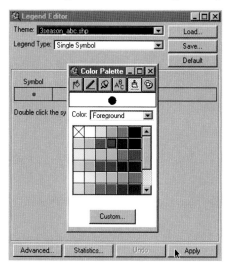

✎ 12a Describe the distribution pattern of 3season_abc.shp.

13 From the Theme menu, click Auto-label. Change the Label field to Site # and click OK.

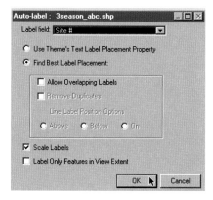

14 Click the Pointer tool. Select and move any labels that are difficult to see on the map.

Now that the sites are mapped and labeled, you are ready to begin mapping the attribute data. Remember, your goal in this exercise is to determine which site or sites will be best for restoring fish habitat. To make that decision, you will look at four variables (water depth, dissolved oxygen, pH, and conductivity). First, you will explore the data on water depth.

15 **Turn off 3season_abc.shp. Make one copy of 3season_abc.shp by using the Edit menu to copy and paste it in the table of contents.**

 16 **Make the top 3season_abc.shp theme active and click the Theme Properties button. Change the theme name to Depth (ft).**

ANALYZE

Water depth

Water depth is an important variable to take into consideration when identifying fish habitat restoration sites. Because the Turtle River is located in northern North Dakota, all or some of the river water will freeze in the winter, killing off fish or forcing them to move to deeper water. Therefore, the only suitable sites for fish habitat restoration are those with water depth of 3 feet or more.

17 **Turn on Depth (ft) and open the Legend Editor. Change the Legend Type to Graduated Symbol and the Classification Field to Depth. Click Classify and choose Natural Breaks and 3 classes. Experiment with symbol colors until the water depth symbols are easily visible in the map.**

17a Which of the sites meet the water depth criteria and what are their depths?

⁒ **HINT:** You can use the Identify tool or the attribute table to determine exact water depth.

18 Close the Identify window or attribute table if they were open. Save your project under a new name (e.g., **water_abc.apr** where "abc" represents your initials). If you are in a classroom environment, ask your instructor how to rename your project and where to save it. Write down the project name and where it is stored.

If time permits, proceed to step 2 of part 2 of the exercise. Otherwise, exit ArcView.

PART 2   SELECTING A SITE FOR FISH HABITAT RESTORATION

1 Start ArcView and open the project you saved in step 18.

In order to examine pH, dissolved oxygen, and conductivity, you will add the fourth season of data to the three seasons you already have.

2 Make the Project window active. Click the Tables icon and click Add. Navigate to your module 4 exercise data and change the List of File Types to Delimited Text (txt). Select 4thseasontable.txt and click OK.

3 Move the 4thseasontable.txt table to the right side of the view. Open the attribute table for 3season_abc.shp and move it to the left of 4thseasontable.txt as in the graphic below.

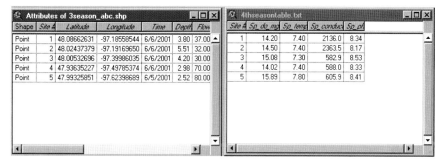

4 In 4thseasontable.txt, click the field name, Site #. Then click the field heading Site # in the Attributes of 3season_abc.shp table.

✳ **NOTE:** Attributes of 3season_abc.shp should be the active window at this point.

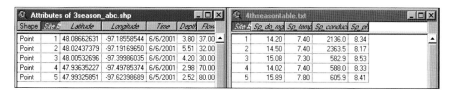

5 Click the Join button. The 4thseasontable.txt table disappears. Scroll across Attributes of 3season_abc.shp and locate the spring data that was just added.

6 Close the attribute table. Return to the view and change the theme name from 3season_abc.shp to **All Seasons**.

In order to thematically map pH, dissolved oxygen, and conductivity, you will need to make three copies of the theme.

7 Copy and paste three copies of All Seasons into your table of contents.

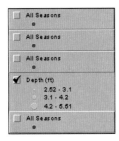

pH

pH is a measure of how much acid is in water. It is measured with a logarithmic scale, where each number on the scale represents a ten-fold increase from the previous number. Normal water has a pH of 7. Acid rain has a pH of 5 and has a hundred times more acid than normal water. Water from areas with limestone may have pH values of 8. It has ten times less (1/10) acid than normal water (also known as basic conditions). pH is tested using litmus or other indicator paper where red shows acid conditions (0–7) and blue-purple shows basic conditions (8–14). Fish need water with a pH of close to 7.0 to survive.

Now you will map the pH for all five sites and all four seasons.

8 Turn on the top All Seasons theme and make it the only active theme. Change its name to **pH**. Turn off Depth (ft).

9 Open the Legend Editor for the pH theme. Change the Legend Type to Chart. In the Fields list, click Su_ph and click Add. Repeat this procedure so that Fa_ph, Wi_ph, and Sp_ph are all added to the right column.

10 Click the Bar Chart icon. Double-click on each of the four season symbol colors to change them to match the picture below. Double-click the Background Symbol and change the symbol color to red. Click Apply.

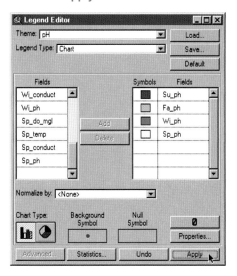

⟱ 10a Name the site(s) with the most acidic water measurement. Name the site(s) with the most basic water measurement.

⟱ 10b Describe the general pattern of pH from summer to spring.

⟱ 10c Based on the information you learned about aquifers (step 5 of part 1), how would you expect the pH to change as the river flows eastward? Why?

⟱ 10d Determine which three sites meet the pH requirements of the fish habitat restoration project (pH average closest to 7.0) and complete a table like the one below. (Do your calculations using a separate sheet of paper. When you use the Identify tool, be sure to click the location of the red background symbol.)

SITE #	Su_ph	Fa_ph	Wi_ph	Sp_ph	AVERAGE pH

Dissolved oxygen

Now you are ready to map the dissolved oxygen data and evaluate the different levels. Dissolved oxygen is a measure of how much oxygen is present in the water. A higher amount of oxygen in the water is conducive to a higher volume of plant and animal life. The fishing club is trying to find an area of the river that consistently has a dissolved oxygen level of at least 10 mg/l year-round. Now you will map dissolved oxygen and evaluate which river sites have the best oxygen concentrations.

11 Close the Identify Results window and turn off pH. Change the name of one of the All Seasons themes to **Dissolved Oxygen** and turn it on. Open its Legend Editor.

12 Change the Legend Type to Chart and add all the "do_mgl" fields. Make sure your selections match the graphic below:

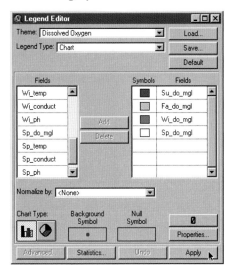

12a What pattern, if any, exists in the dissolved oxygen levels from each site to the next?

12b What trend, if any, exists in the dissolved oxygen levels from one season to the next?

12c Explain the trend identified in 12b.

12d Determine which sites meet the dissolved oxygen requirements of the fish habitat restoration project. Complete a table like the one below.

SITE #	SUMMER	FALL	WINTER	SPRING

13 Close any open attribute tables or Identify Results windows. Turn off Dissolved Oxygen.

Conductivity

In addition to analyzing water depth, pH, and dissolved oxygen, you will evaluate the conductivity levels for each site. The conductivity test is a good indicator of how much salt or ionic compounds are in the water. Some compounds such as table salt (an ionic compound) dissolve in water and break into two charged parts. These charged parts can conduct electricity at the tip of a conductivity probe. Extremely high levels of conductivity can cause dehydration in animal life. Therefore, you will identify the sites on the Turtle River with the lowest levels of conductivity.

14 As you did for pH and dissolved oxygen, change the All Seasons theme name to **Conductivity** and map the season conductivity data using bar charts.

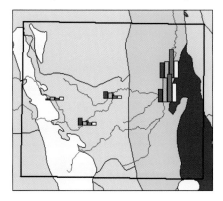

✎ 14a What pattern, if any, exists in the conductivity levels from each site to the next?

✎ 14b Explain the high conductivity levels at sites 1 and 2.

✎ 14c What trend, if any, exists in the conductivity levels from one season to the next?

✎ 14d Determine three sites that meet the conductivity requirements of the fish habitat restoration project. Complete a table like the one below.

SITE #	Su_conduct	Sp_conduct

Now that you have mapped water depth, pH, dissolved oxygen, and conductivity, you are ready to decide where the habitat restoration efforts should be focused.

ACT

15 Turn themes on and off, consult the attribute tables, and consult your answers to question 17a from part 1 and questions 10d, 12d, and 14d from part 2 to determine which sites meet the requirements for the fish habitat restoration project.

✏ 15a Complete a table like the one below by placing an ✕ next to each site number under the requirements it meets.

SITE #	WATER DEPTH	pH	DISSOLVED OXYGEN	CONDUCTIVITY
1				
2				
3				
4				
5				

✏ 15b Based on your completed table, which site or sites would you recommend for fish habitat restoration? Why?

Congratulations. You've analyzed the data and have determined which site(s) would be appropriate for fish habitat restoration on the Turtle River. Now you are able to present your results to the fishing club.

16 Save your project. If you'd like to print a map of the test results, create a layout and print it now. Otherwise, exit ArcView.

SUMMARY

In this exercise, you:

• Added tabular data to the project, and thematically mapped different attributes

• Joined one data table to another

• Symbolized data using bar charts

• Analyzed water-quality data to determine which river sites meet criteria for a fish habitat restoration project

ASK A GEOGRAPHIC QUESTION

Fresh water is arguably our most valuable resource, but it is threatened by many factors, both environmental and human-created. Water supplies are often most affected by the geological makeup of the ground they pass over. We must understand what is in our rivers. We must also understand what outside factors are contributing to their healthy or unhealthy condition if we are to maintain our rivers and streams for years to come. Testing multiple sections of a river or stream over time is an excellent way to conduct a long-term water-quality study of an area.

Developing your geographic question
Your geographic question or questions will be guided naturally by the condition of the river you choose to study and the interest of your organization. The following list represents the wide variety of questions that can be addressed with a water-quality analysis:

- A marked difference in the vegetation along the banks of the river is observed within a short distance. Is the variation in vegetation caused by variation in water quality?
- Why is the fishing great in one stretch of the river and poor in another?
- How does the presence of a golf course along one stretch of the river affect water quality downstream?
- Does water quality differ between the sections of the river upstream from an urban center and downstream from it?
- Many local companies run river-rafting trips down the same stretch of the river. What effect, if any, does this have on water quality?
- Construction of a major housing development is about to begin near the river. How does the quality of the water compare before, during, and after the construction?

Whatever your geographic question, you will need to collect data over a period of time. Some questions will be best answered if data is collected daily. In the case of the Red River High School study, the students collected data once during each of the four seasons. Once you've settled on a geographic question, be sure to define a test period.

Other natural places to conduct similar studies of change over time

➤ Open fields
➤ Tree rows
➤ City parks
➤ Empty lots

Issues to consider when selecting testing sites

Before you begin your quest to evaluate the health of your river, decide what sections of the river you want to study. When deciding the geographic extent of your study, consider several factors, such as your time and budget, whether or not there is a particular place on the river that is of interest or problematic, and the accessibility of the reach of the river you want to study.

ISSUE	CONSIDERATIONS
Access	• Is the access to the river site through public or private property? • Bridges tend to provide access paths to rivers through public property. • Will the depth of water (either too shallow or too deep) affect your ability to collect water samples at each site? • Will you be able to access the water samples during each testing period? (For instance, if you're testing seasonally, you may need to drill a hole through ice in the winter to get access to liquid water.)
Safety	• Make sure all people collecting water know how to swim. • The depth and swiftness of the river should be considered when selecting testing sites. Don't select sites that will be dangerous to water testers. • Be aware of possible downstream hazards for each site (e.g., waterfalls, dams, white-water rapids) and select your sites accordingly. • If river water is extremely cold, protect your water testers against hypothermia by outfitting them with appropriate insulated waterproof gear. • Conduct a safety orientation for all on-site water testers before your test period begins.
Location of sites	• Each testing site should be within a reasonable driving distance from the lab or location where you will perform the chemical tests. • Will you have transportation for all water testers (e.g., vans, personal vehicles)?

ACQUIRE GEOGRAPHIC RESOURCES

The geographic resources you need to complete a water-quality analysis can be divided into three categories: basemap data, water-quality data, and community partnerships. Basemap data includes the general data, such as location of rivers, roads, aquifers, land use, and so on, that provides a geographic context. Water-quality data is the data that you will collect at each testing site. Different organizations may be able to help you find geographic resources or help you interpret your geographic data. You should consider forming partnerships with these organizations.

Basemap data

One influence on the quality of water found in different rivers is the surface geology of the area. You can research the local geology, including glacial history, the presence of particular types of aquifers, and soil types that may affect the river water. In order to analyze the water-quality data you collect, it may be helpful to have the geologic information in your GIS. The following table provides a list of data you will need for your basemap and suggestions for places to look for it:

TYPE OF DATA	WHY YOU NEED THE DATA	WHERE TO LOCATE THE DATA
Rivers	Provides location of the river of study	• ESRI Data & Maps Media Kit • Your state departments of natural resources and fish and wildlife • Your town or county planning department • National Hydrography Data Set at *nhd.usgs.gov* • Digitize from an aerial photograph or USGS map (refer to module 8)
Roads	Shows where roads are in relation to the river	• ESRI Data & Maps Media Kit • Your state departments of natural resources and fish and wildlife • Your town or county planning department • U.S. Census 2000 TIGER/Line data from *www.geographynetwork.com*
Aquifers	Shows locations and characteristics of aquifers	• *www.nationalatlas.gov* • Your state departments of natural resources and fish and wildlife • Your town or county planning department
Watershed	Watersheds are biologically distinct units. The watershed boundary for the river you are studying provides a logical map extent or study area boundary.	• ESRI Data & Maps Media Kit • Your state departments of natural resources and fish and wildlife • Your town or county planning department • National Hydrography Data Set at *nhd.usgs.gov*
Study area	Provides an easy way to zoom in on the area you are studying	• Extract the area from another theme (e.g., if your study area is a city or park) • If your study area is unique, digitize the boundary by using other themes as a reference
Digital	Provides a bird's-eye view of the surrounding geography of the sites you are sampling	• Your state departments of natural resources and fish and wildlife • Your town or county GIS or planning department • Two excellent sources for DOQs online are *www.usgs.gov* and *www.terraserver.microsoft.com*

Water-quality data

As you research the geologic history of the river you are testing, develop a list of variables to test. For example, if you learn that part of your river flows over a limestone aquifer, then make sure you test carbonate levels because limestone greatly influences them. Your geographic question will also direct the list of variables you test. For example, if you are trying to determine how a golf course affects the water quality of a nearby river, you will most likely test for nitrate levels because nitrates are in abundance in the fertilizers used on golf courses or agricultural lands.

The following table lists some variables you could test in your water-quality study.

VARIABLE NAME	VARIABLE DEFINITION	UNITS OF MEASURE
Temperature	Measure of average kinetic energy (speed) of atomic particles. (The skin senses this as warmth or cold.)	Centigrade or Fahrenheit
Depth	Depth of the river at its deepest point.	Feet or meters
Flow	Distance the water moves in one minute in the center of the stream or river.	Feet per minute
Turbidity	Clarity of water; usually tested with a turbidity disk (Secchi disk).	Centimeters of depth at which you can still determine the black-and-white pattern of a Secchi disk
Dissolved oxygen (DO)	Needed by aquatic animals for cellular respiration. Colder water and waters with moderate amounts of plant life or areas of fast-running water will be high in oxygen.	Milligrams per liter (mg/l) or parts per million (ppm)
Conductivity	Amount of electricity that is able to flow through a set distance of sampled river water. It measures how much salts (ionic compounds) are dissolved in the water.	Microsiemens (μS)
pH	Measure of how much acid is in the water (the hydrogen ion concentration). Normal healthy water is pH 7. Acid conditions are pH 1–6 and basic conditions are pH 8–14.	There are no units. The scale is the log of the hydrogen concentration in moles per liter.
Calcium, magnesium, iron	Presence of these elements causes water to become "hard" (sticky clean when you wash yourself).	Parts per million (ppm) or milligrams of that element per liter of water (mg/l)
Ammonium	Ammonium nitrate is a common ingredient in fertilizers.	Parts per million (ppm)
Carbonate	Common component of limestone/karst rock formations; causes rivers to have basic pH.	Parts per million (ppm) or milligrams per liter (mg/l)
Phosphate, nitrate	Commonly used as plant nutrients and found in fertilizers.	Parts per million (ppm)
Sulfate	Common ion originating from natural geologic sources. It can cause scale buildup in pipes and boilers.	Parts per million (ppm)
Fecal coliform	Number of noninfectious bacteria existing in waterways.	Number of colonies per culture

After creating your list of variables to be tested, it's important to have all the testing equipment needed for each test. The following list contains tips on where to obtain water-testing kits and equipment:

EQUIPMENT NEEDED	WHERE TO FIND IT
Electronics	• Graphing calculators, hand-held computers, and laptops with probes can be purchased through online retailers, major electronics stores, or mail-order catalogs. • Investigate the possibility of borrowing electronic equipment from a local partner.
Testing probes	• Water-testing probes can be purchased through online retailers, major electronics stores, or mail-order catalogs. • Other sources include forestry and science educational supply companies. • Investigate the possibility of borrowing probes from a local partner.
Chemical-testing kits	• A variety of different chemical-testing kits is available online or via mail order from science and educational supply companies. • Look at different kit types to ensure the kit you purchase includes enough tests for your project.
GPS units	• Hand-held GPS units can be purchased through online retailers, major electronics stores, sporting goods stores, or mail-order catalogs. • Purchase GPS units that collect coordinates in decimal degrees. • Consider purchasing GPS units that have some protection against water (e.g., a carrying case or waterproofing). • Investigate the possibility of borrowing GPS units from a local partner.
Waders	• Overalls, thigh boots that hook to a belt, and knee-high boots can be purchased through online retailers, outdoor stores, or mail-order catalogs. • Waders made of neoprene material will be the warmest. If warmth isn't a factor in your area, purchase canvas or vulcanized products. • Consider the depth of the water you'll be testing before making these purchases. Knee-high boots will be sufficient only for water up to 1 foot deep.
Bottles	• Polyethylene water bottles of varying sizes are available online or via mail order from science and educational supply companies. • You can use regular polyethylene soda bottles that have been cleaned and rinsed with distilled water.

Community partnerships
At this point, you may be concerned that you do not have enough knowledge of chemistry or the geology of your area to embark upon a water-quality study. This should not be an obstacle because there are many experts in those fields who would be good resources for your project and who would be glad to share their expertise. In the project-planning stages, consider partnering with a variety of organizations. Quite often, they will be interested in your work and will be able to provide expertise or necessary equipment.

POSSIBLE PARTNER	BENEFITS OF PARTNERSHIP
Chemistry or geology teacher in your local school district or professor from a local college or university	• Has valuable technical knowledge of chemistry and geology concepts and most likely has knowledge of local region • May provide students who could help with testing • May provide a lab environment for performing water tests
Soil or local water extension agent	• Has in-depth experience and knowledge of local soils and rivers • May have equipment you can borrow • May know of other agencies or people that you can borrow equipment from or partner with • May be able to connect you with a water-testing lab to do the water analysis
City water-testing lab	• Has in-depth knowledge of chemical makeup of the local watershed • Has resources on state standards for water testing • May allow you to borrow water-testing kits or equipment • Could provide a lab environment for performing water tests
Local conservation organizations	• Knowledgeable about local region and aware of various projects and groups doing similar work • Could provide volunteer water testers

EXPLORE GEOGRAPHIC DATA

Technical issues

There are several issues to consider when using GIS to study water quality.

• Create a data table using a spreadsheet program or text editor. To thematically map your sample sites and data using GIS software, set up your table so the sample sites are in the rows and the chemical variables are in the columns. Refer to the example below:

SITE #	1_Temp	1_Flow	1_Depth
1			
2			
3			

• When adding new data that was taken during one time period to data from another period, join the new data tables to your initial table. Set up one common field in your table such as Site#. Make sure the contents of the field have the same format, such as "string" or "numeral." Also, make sure all the field names indicate in which time period the data was collected. Refer to the example below:

SITE #	1_Temp	1_Flow	1_Depth	2_Temp	2_Flow	2_Depth
1						
2						
3						

120

Preliminary explorations

Before beginning your actual analysis of the data, take time to explore the data and familiarize yourself with its content, characteristics, and potential spatial patterns. Research the normal ranges of the variables to determine whether your river water is in a healthy range. You can find this information on chemical supply Web sites, in water-quality handbooks, or in a good biology textbook that deals with water quality.

Suggestions for preliminary data exploration:

- Look at your data table to be sure that all the data is in the correct units. For example, pH ranges should be from 0 to14, with most close to 7. If there are numbers outside this range, there was probably an error in data entry or malfunctioning equipment.

- Use the Legend Editor to symbolize the water-quality data in a variety of ways: graduated color for pH, chart for conductivity, and so forth.

- Reorder themes in the table of contents so you can see how the basemap data and water-quality data overlay.

- Note visual patterns on how each variable changes with each testing site.

ANALYZE GEOGRAPHIC DATA

When analyzing your data, try to identify the preliminary questions that you will need to answer before you can answer the broader, essential question of your study. For example, if your ultimate question is "What areas of the river are most suitable for fish habitat restoration?", you might want to identify and explore the following preliminary questions:

1 Where are the testing sites on the river?

- Add GPS location data for each testing site.
- Add basemap data including river location, geologic history, and land use.
- Symbolize and label the testing sites.

2 Which sites are deep enough for fish habitat?

- Add collected water-quality data to the project.
- Thematically map water-depth data.
- Look at the attribute table to determine which sites meet the requirement.

3 Which sites meet the chemical-testing requirements for fish habitat?

- Add collected water-quality data to the project.
- Thematically map and symbolize the various attribute data.
- Use visual analysis, the Identify tool, or theme-on-theme selection to determine which sites meet the different requirements.

Drawing conclusions

When concluding your study of a local river, you should refer back to your original geographic question. Review the variables you tested and develop your conclusions accordingly. In the Turtle River case study, the following factors were reviewed and "checked off" to see what parts of the river had the best conditions for fish habitat:

- Water with minimum flow and depth requirements
- Must support life (have enough oxygen, maintain a moderately cool temperature, have clear enough water to support plants' need for sunlight)
- Be in the right pH range
- Must not be overly fertilized (with high nitrate and phosphate levels)

After the sites were identified, the students were ready to act on their results.

ACT ON GEOGRAPHIC KNOWLEDGE

After analysis, it's time to put this valuable information to use. The information you have collected could be vital to public officials for preserving or improving water quality in the area, educating the community, and making decisions regarding use of the river. It is an essential instructional tool to teach people about one of the most important natural resources in their community.

Possible action steps

- As part of a community education program, prepare a presentation to the public that explains your results and any recommendations you have for the future of the local river.
- Continue public outreach with an article in the local newspaper and advertising through flyers and billboards.
- Convince local television stations to do a spot on your group's involvement with the river.
- Become part of local or state park interpretive programs.
- Present your recommendations to resource management professionals such as the Department of Natural Resources.
- Approach economic development groups or local government agencies with your findings to encourage renewable use of local rivers such as ecotourism or outdoor sports.
- Educate others by volunteering at a local school to present your findings to students in the classroom.
- Look into federal and private grant programs that provide money to support water research and conservation initiatives at the local level.

Next step

Even though you have completed the initial study, continuing to monitor the testing sites could result in meaningful and useful information. Develop a plan for continued analysis. You might want to watch changes in land use or habitat development near the river to see the type of impact they have on the water quality.

In addition, form an alliance with other groups doing water-quality studies on nearby rivers. You can share resources, data, and information to develop a big picture of the water quality in the watershed. The work you have put into studying your river is an invaluable resource to your community and the surrounding communities.

MODULE 4 ACKNOWLEDGMENTS

Thanks to the students of Red River High School and Roger Palmer for providing us with this case study.

The Turtle River bottom profile map was provided courtesy of Red River High School and is used with permission.

Water quality and Turtle River location data provided courtesy of Red River High School and used with permission.

Aquifer data provided courtesy of North Dakota Water Commission.

During the twentieth century and up to the present day, landfills have been the principal means of waste disposal in the United States and the rest of the industrialized world. Originally located outside of population centers, many former landfill sites today are dangerously close to residential and recreational areas. Such sites, built before tougher environmental standards were mandated, represent potential hazards to nearby populations and environments. Toxic chemicals and gasses leach into nearby land and water sources, harming the natural environment.

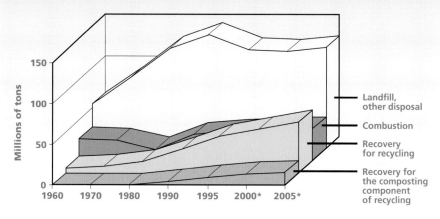

Landfill, other disposal
Combustion
Recovery for recycling
Recovery for the composting component of recycling

Millions of tons

*Assumes 30% recovery in 2000 and 32% recovery in 2005.
Source: Characterization of MSW in the U.S.: 1998 Update, U.S. EPA, Washington, D.C.

CASE STUDY

Identifying potentially harmful landfills

Students at the Crescent School in Toronto, Ontario, learned that there were nearly seventy abandoned landfill sites in the greater Toronto region. After acquiring data about these sites from a local environmental organization, the students used GIS to map the sites and analyze their proximity to area rivers, schools, hospitals, and parks. Their analysis demonstrated the power of GIS to identify potential environmental hazards and guide decision makers in implementing appropriate mitigation measures.

EXERCISE

Map, query, and analyze neighborhood data to identify high-risk landfills
You will use GIS to identify potentially hazardous landfill sites in one Toronto
neighborhood. After locating landfill sites on a neighborhood basemap, you
will analyze patterns of land use and population near those sites. By exploring the proximity of the sites to rivers, schools, and residential areas, you
will identify neighborhood sites that should be given the highest priority for
further environmental impact studies.

ON YOUR OWN

Explore the resources and data needed to identify and analyze potentially
hazardous point-source pollution sites in your own community. This section
will provide suggestions for enlisting the support of local environmental
groups and data providers, data exploration and analysis, and ways to act
on your study's findings.

Identifying potentially harmful landfills

Greater Toronto area, Ontario, Canada
Closing a landfill and covering it with a park or a housing project may once have seemed like a good idea, but now, as much as fifty years later, some residents of the greater Toronto area are concerned about the landfills in their neighborhood. Nearly seventy abandoned landfill sites dot the region, and studies have shown that such landfill sites can lead to a variety of ill effects on the environment and on human life. Concerned eleventh- and twelfth-grade students in Rex Taylor's World Issues class at Crescent School in Toronto, Ontario, conducted a GIS analysis of the landfill sites and increased the awareness of this public health threat by presenting their findings to local environmental groups.

THE GEOGRAPHY OF WASTE DISPOSAL—A LOCAL AND GLOBAL ISSUE

High school students in the World Issues class use geotechnologies such as ArcView software to study a local problem or issue with global implications. They were offered the opportunity to work in small groups to do a major semester-long assignment. At the time, the Toronto news media were pursuing the story of the "Walkerton Disaster." In this highly publicized case, seven people died and hundreds were sickened by well water contaminated with bacteria from a nearby intensive livestock operation.

A group of students brainstormed possible research topics related to the disaster. Not only did they need a compelling topic, they also needed to be able to obtain data they could analyze with GIS. They came up with the idea of old dump sites, another pollution threat to water resources. A quick search of the Internet proved the students had chosen another important local issue.

Courtesy Lake Ontario Keeper

The abandoned Beechwood Drive landfill in East York, Ontario, appears as an open meadow. Old abandoned landfills like this one were often buried, forgotten, and later turned into residential areas, public parks, or school playgrounds.

Students learned that people in southern Ontario are becoming increasingly aware of the public health implications of dozens of old landfills and industrial and agricultural waste sites throughout the region. People are worried because some landfill sites are near rivers or lakes—and these are water sources tapped for households across Ontario. To add to this, scientists suspect that in rural Canada many household wells are contaminated by substances from such common sources as septic systems, fertilizer, pesticides, and livestock wastes.

Using landfill data from the Internet and basemap data already available at their school, students mapped the locations of abandoned landfill sites in and around Toronto (red dots). When active, these landfills were typically at the outskirts of town, but because of urban growth, the sites are now well within populated areas. Students were surprised to learn that five different landfill sites are located within 4.5 kilometers of Crescent School.

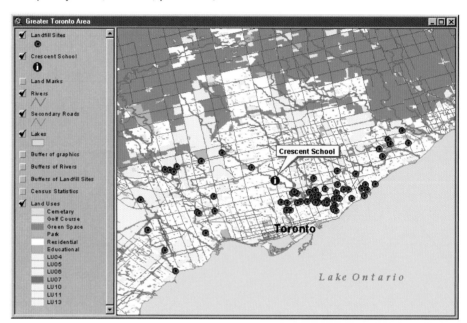

In addition to substances that move through the soil and water, landfills produce naturally occurring gases such as methane and carbon dioxide. The gasses form inside the landfill as the waste decomposes. Some of them escape through the surrounding soil or simply move upward into the atmosphere, where they drift away and may move into the communities. Depending on the wind and weather, these gasses may cause odors in neighborhoods.

Modern landfills in Canada and the United States are typically built to exacting standards that include liners to prevent soil and groundwater contamination, methane gas collection and dispersal systems, and regulations preventing disposal of hazardous waste into the nearby environment. But old landfills were not covered by these measures.

Studies of industrial dumps and contaminated water supplies during the last decade have reported numerous negative health effects on the human population. In Hamilton, Ontario, a study revealed significantly elevated rates of a number of conditions in people who lived or worked near an industrial dump compared to people living farther from the site. These conditions included bronchitis, difficulty breathing, cough, skin rash, arthritis, and heart problems (Hertzman et al. 1987).

A recent government-funded study in Britain found a 1-percent increase in the risk of birth defects to babies born near a landfill site. This risk increases to 7 percent for those near a hazardous waste site (Elliott et al. 2001).

Courtesy Lake Ontario Keeper

Courtesy Lake Ontario Keeper

The Don River in East York, Ontario, looks pure and idyllic in the scene at the left. In fact, it runs past the foot of the old Beechwood Drive landfill. Pollutants seep into the river (right), which is part of a major drainage system for the region that empties into Lake Ontario near downtown Toronto.

These examples show that a landfill can have an adverse effect on the health of the environment surrounding it. The students wondered whether or not there was a pattern in the location of the abandoned landfills in the greater Toronto region, recognizing that such a pattern could be important to public health and welfare. They also wanted to know which, if any, of the landfills pose a threat to public places, particularly those in their own community, or to the rivers and streams flowing into Lake Ontario. They decided to use GIS to locate and map the landfills and to investigate the answers to their questions.

DIGGING DEEPER

The students' initial Internet search led them to the Web site of Lake Ontario Keeper, a local environmental organization. The group is concerned with the purification and revitalization of natural water sources around Lake Ontario, including those in the Toronto area. In addition to a wealth of background information about the facts and issues surrounding abandoned landfills, the Web site contained the data students were looking for.

The next step of the three-week investigation was to get the data into ArcView and map the landfill locations. Students created ArcView 3.2a projects with basemap themes from the school's existing GIS database. These themes included streets, census data, landmarks, land use, lakes, and rivers. Using the streets theme and the tabular data from the Internet, students created a shapefile with points locating and identifying each landfill site.

Lake Ontario Keeper

Municipality	East York
Lot or Street No	End of Beechwood Dr.
Class of Danger	A5
Date Closed	1966
Site No from Ontario Ministry of Environment inventory	3031

Courtesy Lake Ontario Keeper

The Lake Ontario Keeper Web site offered students the landfill data they needed for their GIS project—a table like the one above for each of the sixty-seven abandoned landfills in the Toronto area. The table includes the site's location by municipality and street intersection, the danger level of the landfill according to the Waste Disposal Site Classification System of the Ontario Ministry of Environment, and the year the site was closed. The Beechwood Drive landfill is classified as an A5 site, the most dangerous to humans.

After discovering Lake Ontario Keeper on the Web, students contacted the organization. The group was eager to assist in any way they could and agreed to help students present their findings to municipal government officials.

A variety of geographic information was explored to determine its relevance with respect to the landfill data. Maps of four major areas of the greater Toronto area were created, displaying landmarks such as schools, hospitals, residential areas, commercial areas, industrial areas, rivers, lakes, and landfill sites.

The students used keywords such as the names of local cities or regions, "closed landfill sites," and "old landfill sites" when searching for data and information on the Internet.

Census data at different levels of geography (enumeration areas, census tracts, and so on) was compared to the landfill locations. Data on income and population was scrutinized most thoroughly because the students hypothesized that the more economically influential communities were able to affect placement of these landfills. They also expected to find that older communities with higher population densities have fewer landfills.

Next, the students created buffers around features of interest such as schools, hospitals, and rivers, and then calculated the number of landfill sites within certain distances of particular features. Finally, students wanted to more closely examine some of the sites. Because they wanted to minimize the amount of hard-to-get large-scale data, students decided to focus their in-depth research on East York, the community where Crescent School is located. The school is situated near the West Fork of the Don River and several landfill sites are nearby.

RESULTS CONFIRM CAUSE FOR CONCERN

The students found that a majority of the old landfills in the greater Toronto area should be considered potentially dangerous sites according to their criteria. They determined that nearly two-thirds of the abandoned sites are within 500 meters of a river, with more than half within 500 meters of a school. One public park was notably threatened: four different abandoned landfills are located on the edge of Rowntree Mills Park at Finch Street and Weston Road, with another landfill site only 500 meters away.

130   Community Geography: GIS in Action

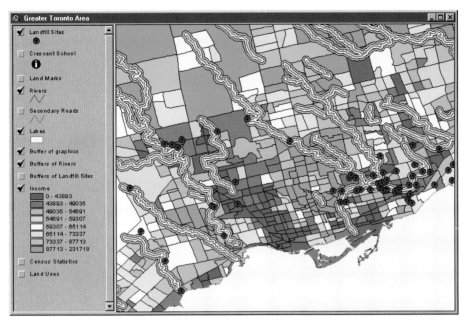

The students' hypothesis about income levels proved to be correct: significantly more of the sites are located in areas of lower income (purple) than areas of higher income (green). The landfills also tend to be located in the vicinity of rivers, as revealed by the quarter-kilometer (white) and half-kilometer (gray) buffers in this map.

When students conducted an in-depth analysis of the twenty-three landfill sites in their own community of East York, they found even higher rates. For example, all of the sites in East York are close to public parks, and most are also near residential areas.

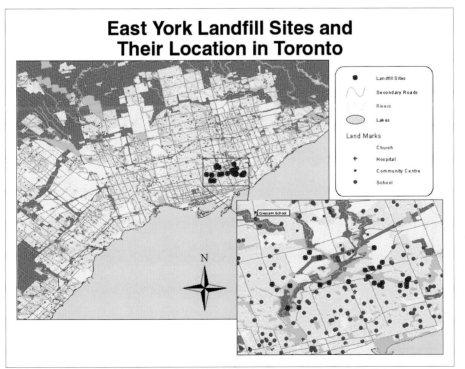

East York Landfill Sites and Their Location in Toronto

In East York, Ontario, 18 of 23 (78 percent) of the landfill sites are located within 500 meters of a river, and 57 percent are within 500 meters of a school. All sites are within 500 meters of a park, and 91 percent are within 500 meters of a residential area.

SPREADING THE WORD

In April 2002, the students presented their finished project to representatives of Lake Ontario Keeper and the Friends of the Don, another organization focused on water-quality issues in the Toronto region. The presentation included both maps and a slide show. The presentation spurred interest within the Lake Ontario Keeper organization to better understand the importance of GIS in mapping the locations of landfills.

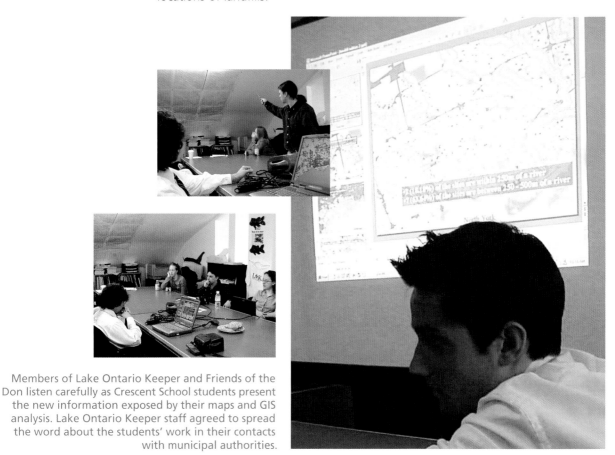

Members of Lake Ontario Keeper and Friends of the Don listen carefully as Crescent School students present the new information exposed by their maps and GIS analysis. Lake Ontario Keeper staff agreed to spread the word about the students' work in their contacts with municipal authorities.

NEXT STEPS

Lake Ontario Keeper staff plan to visit Crescent School to learn more about how the students used ArcView software and to find out how GIS could aid them in their own research. Two of the five students that worked on the project will be returning to Crescent School for their graduating year. They intend to expand the study of landfill sites and to continue working with Lake Ontario Keeper on ways to increase public awareness of this issue.

SUMMARY

ASK A GEOGRAPHIC QUESTION	• Is there any pattern in the location of old, abandoned landfill sites across the Greater Toronto region? • What level of danger do these landfills pose to public places in the surrounding region? • What level of danger do these landfills pose to the region's environment and water resources?
ACQUIRE GEOGRAPHIC RESOURCES	• Obtain abandoned landfill data including landfill locations by street intersection or GPS coordinates and city landfill attributes. • Obtain local-area basemap data such as local street data for mapping locations, major streets, rivers, lakes, landmarks (schools, hospitals, and so on), land use, and census area and statistics.
EXPLORE GEOGRAPHIC DATA	• Create views and symbolize themes. • Locate landfills using latitude and longitude. • Visually analyze landfill distribution patterns.
ANALYZE GEOGRAPHIC INFORMATION	• Calculate buffers around rivers and landfills. • Compute statistics on the number of landfills near rivers and the number of schools (or public places) near landfills. • Perform detailed analysis on a subset of the study area.
ACT ON GEOGRAPHIC KNOWLEDGE	• Create maps and layouts communicating the results of the investigation. • Prepare a slide presentation about the project and results. • Present results to representatives of local environmental organizations. • Plan continued research for next year.

Map, query, and analyze neighborhood data to identify high-risk landfills

Exercise

In this exercise, you will create a map depicting geographic features that provide locational reference, also known as a basemap. You will explore the basemap features of land use, rivers, roads, and the East York neighborhood boundary. In order to analyze high-risk landfills, you will map and query the landfill data. Theme-on-theme selection allows you to select landfills within a specific distance from other features like rivers and schools.

The ⊕ icon indicates questions to be answered. Write your answers on a separate sheet of paper.

ASK

A local environmental group has asked you to evaluate the danger of the twenty-eight abandoned landfill sites in Toronto's East York neighborhood. The group would like you to identify the specific sites that pose the greatest risk to the neighborhood's environment and its population. The environmental organization has defined high-risk landfills as those located inside residential areas and within 0.25 kilometer of a river or school.

PART 1   CREATE AND EXPLORE A BASEMAP

1   Start ArcView and create a new project with a new view.

ACQUIRE

2   Click the Add Theme button and navigate to the module 5 exercise folder *(C:\esri\comgeo\ module5)*. Hold down the Shift key and select the following themes: boundary.shp, landuse.shp, rivers.shp, and sec_roads.shp. Click OK.

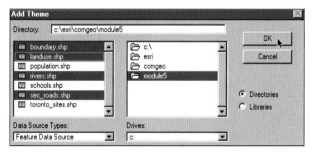

Before continuing with your analysis, you will take a few moments to organize and symbolize the new themes. If you're using ArcView 3.0a, you will need to load the legends for the new themes. If you're using ArcView 3.1 or higher, they were loaded automatically, so skip to step 4.

3 For each of the four themes, open the Legend Editor and click the Load button. Load the appropriate legend file from the exercise folder. For example, for the boundary.shp theme, load boundary.avl. Remember to apply the changes. Close the Legend Editor when you are finished.

4 Make each theme active and use the Theme Properties button to change all of their names as shown below:

OLD THEME NAME	NEW THEME NAME
Boundary.shp	East York
Sec_roads.shp	Major Roads
Rivers.shp	Rivers
Landuse.shp	Land Use

5 Reorder the themes in the table of contents so they match the graphic below. Make your table of contents and View window larger.

6 Turn on all the themes. Make East York the active theme and click the Zoom to Active Theme button.

7 Use the View Properties to name the view East York Neighborhood. If necessary, change the map units to decimal degrees and the distance units to kilometers. Click OK. Your map view should now look like the graphic below.

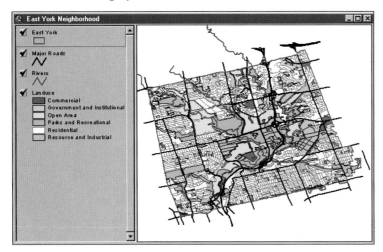

EXPLORE

8 Use the Measure Tool to determine the approximate dimensions of the East York neighborhood.

8a How long is the east–west axis?

8b How long is the north–south axis?

9 Observe relationships between land use and other features of the East York neighborhood.

9a What is the predominant land use in the neighborhood?

9b What, if any, land-use patterns are associated with Rivers?

9c What, if any, land-use patterns are associated with Major Roads?

Prior to investigating where landfills are, you will add school and population data, then explore where people live in the East York neighborhood.

10 Turn off Land Use. Click the Add Theme button and navigate to the module 5 folder *(C:\esri \comgeo\module5)*. Add schools.shp.

11 Change the theme name Schools.shp to **Schools**. Turn on Schools.

12 Open the Schools Legend Editor. Change the symbol to a solid blue star and make the symbol size **14**. Apply your changes and close the Legend Editor.

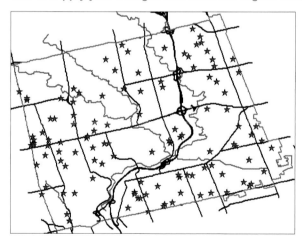

12a How many schools are there in the East York neighborhood?

❊ **HINT:** Open the Attributes of Schools table and look at the number of records indicated on the toolbar.

13 Turn off the Schools theme and close the Attributes of Schools table. Add population.shp from the module 5 folder *(C:\esri\comgeo\module5)*.

14 Drag Population.shp to the bottom of the table of contents. Change its name to **People/sq. km** and turn it on.

 ⁘ **NOTE:** If you are using ArcView 3.0a, you need to load the legend file for Population.shp. Refer to step 3 for detailed instructions.

15 Turn the Land Use theme on and off to observe the relationship between it and the People/sq. km theme.

15a Which areas of East York are densely populated (more than 5,000 people per square kilometer)? Describe the location of specific areas by referring to major roads by name. For example, "There is relatively high population density along Victoria Park Avenue in the eastern part of the neighborhood." Describe two more areas.

15b Give two possible reasons why some residential sections might have a higher population density than others.

15c Is East York more densely populated in the northern half or the southern half of the neighborhood?

16 Close the Identify Results window if it is still open. The basemap is now complete and you should save your project. If you are in a classroom environment, ask your instructor where to save your project and how to rename it. Write down your project name and where it is stored.

17 If necessary, exit ArcView. Otherwise, proceed to step 2 of part 2.

PART 2　EXPLORE AND ANALYZE LANDFILL DATA

1 Start ArcView and navigate to the location where you saved your project from part 1.

2 Use the Add Theme button to add landfill data from the module 5 exercise folder *(C:\esri \comgeo\module5)*. Select Toronto_sites.shp from the list.

3 Rename the Toronto_sites.shp theme **Landfills** and turn it on. Open the attribute table for Landfills and, if necessary, reposition it so you can see both the map and the table at the same time.

 ⁘ **NOTE:** If you are using ArcView 3.0a, you need to load the legend file for Toronto_sites.shp. Refer to step 3 of part 1 for detailed instructions.

3a How many landfills are located in East York?

3b Are there more landfill sites in the northern half or the southern half of the neighborhood?

3c Based on your answer above (3b) and your answer to question 15c in part 1 about East York's population density, what generalization could you make about the difference between the northern and southern parts of the neighborhood?

✷ **NOTES ON FIELD DEFINITIONS:**

Latitude and longitude
Because the latitude and longitude coordinates are rounded, they are not sufficiently precise to discriminate between small distances reflected in different street addresses. For this reason, two or more sites may be listed as having the same latitude and longitude coordinates.

Class
This field refers to the "Danger Class" of the site. This classification, which is intended to be an indicator of the site's potential threat, is based on a system developed by the Ontario Ministry of Environment. A3 means that the site is deemed to have the potential to affect human health, is composed of household waste, is in an urban area, and has been closed less than ten years. A5 sites are the same except that they have been closed ten to twenty years.

3d Which category do you think is likely to present the greater danger, A3 or A5? Explain.

Now you will explore the landfill data to identify how many landfills are close to schools and rivers.

4 Make Landfills the active theme. From the Theme menu, choose Select By Theme. First, choose "the selected features of" Rivers from the second pull-down list. Then choose "Select features of active themes that" Are Within Distance Of from the first pull-down list. Type **0.5 km** for Selection distance. Make sure your Select By Theme window matches the graphic below, and then click New Set.

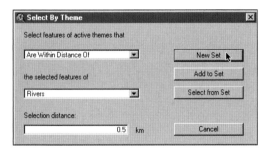

All the landfills that are within 0.5 kilometer of a river are selected in the view.

4a How many of East York's twenty-eight landfills are within 0.5 kilometer of a river?

5 Clear the selected features and use the theme-on-theme selection procedure you learned in the previous step to answer the following questions:

5a How many of East York's twenty-eight landfills are within 0.5 kilometer of a school?

5b How many of East York's 119 schools are within 0.5 kilometer of a landfill?

✷ **HINT:** Be sure schools is the active theme.

6 Clear the selected features. Save your project.

PART 3   IDENTIFY AND MAP HIGH-RISK LANDFILLS

The task that you have been given is to identify the landfill sites that represent the greatest potential risk to the people and environment of East York. These have been defined as landfill sites that are located within residential areas and are also within 0.25 kilometer of rivers or schools. First, you will select all the landfills that are close to rivers and schools. Then you will query the selected data to select landfills located in residential areas.

ANALYZE

1   Make Landfills the active theme. Use the procedure described in part 2, step 4 to select all landfill sites that are within 0.25 kilometer of a river. Click New Set.

All of the landfills within 0.25 kilometer of a river are selected. In this map, they are yellow.

2   With Landfills as the active theme, choose Select by Theme from the Theme menu. Select all the landfills that are within 0.25 kilometer of a school. Click Add to Set.

Now, the selected landfills on your map represent landfills that are within 0.25 kilometer of a river or school.

✳ **NOTE:** Do not clear the selected sites.

✎ 2a How many of East York's twenty-eight landfills are very close to rivers or schools?

Now that you have selected all the landfills that are close to rivers or schools, you are ready to narrow your search further. You will query your land-use data for all residential areas and then perform theme-on-theme selection to determine which selected landfills are in residential areas.

3 Make Land Use the active theme and then turn it on. Click the Query Builder button. In the Fields list, double-click [Category], single-click the equals sign (=), and from the Values list, double-click "Residential." Make sure your Query Builder window matches the graphic below and click New Set.

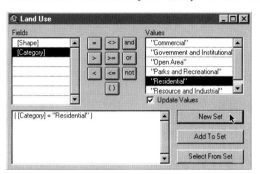

4 Close the Query Builder window. Make Landfills the active theme again. Perform theme-on-theme selection to find landfills that are completely within the selected features of Land Use. Click Select From Set.

The selected landfill sites meet the criteria you were given to define high-risk sites: they are all within 0.25 kilometer of a river or school and are located within residential areas.

5 Make Land Use the active theme and clear the selected residential areas to better see the selected landfill sites.

5a Based on this analysis, how many of East York's twenty-eight landfills are designated high risk?

5b Is there any spatial pattern in the location of the selected sites? If so, describe it.

5c Which river is in the greatest danger from these high-risk sites?

5d Identify the schools that are within 0.25 kilometer of any high-risk landfill.

6 Zoom to the East York theme.

7 Make Landfills the active theme. From the Theme menu, select Convert to Shapefile. Navigate to the directory where you would like to save the new file. If you are in a classroom environment, ask your instructor for directions on where to save it. Name the new file **highrisk_abc.shp** where "abc" represents your initials. Add the new theme to the view.

8 Turn on Highrisk_abc.shp. Rename the theme **High Risk Landfill Sites**.

9 Open the Legend Editor for High Risk Landfill Sites. In order to distinguish the A5 sites from the A3 sites, select Unique Value for Legend Type and select Class for the Values field. Choose a size and color for each symbol that will show clearly on your map. Click Apply and close the Legend Editor.

10 Make Landfills the active theme and click the Clear Selected Features button.

Now that all of the high-risk landfills are symbolized, you are ready to prepare a presentation layout to give to the environmental organization.

ACT

11 From the View menu, choose Layout. Select the Landscape template and click OK. Resize the Layout window by dragging the lower right corner to make it larger.

12 Double-click the scale bar and change the units to kilometers.

T 13 Use the Text tool and click the layout where you want your name and date to go. Complete the Text properties window and click OK.

14 Look over your presentation layout and make final changes. Your layout is now ready to print.

15 Print your map.

16 Save your project. Exit ArcView by choosing Exit from the File menu.

SUMMARY

In this exercise, you:
- Added community and abandoned landfill data and thematically mapped attributes
- Observed spatial patterns of community and landfill data
- Used theme-on-theme selection and the Query Builder to identify landfills that were close to schools or rivers and in residential areas

ASK A GEOGRAPHIC QUESTION

Humankind has always used our planet as a "waste sink," meaning that Earth must absorb all the waste products of human activities. However, in recent centuries, population growth and industrialization have strained Earth's ability to fulfill this role while maintaining sustainable systems.

Virtually every community faces potential danger from point-source pollution—environmental damage emanating from a specific location or point. Examples of such pollution sources include abandoned landfills, ground seepage from former manufacturing sites, leaking underground gas tanks, or untreated industrial or sewer discharge. GIS is an ideal tool for observing, exploring, and analyzing the impact of such pollution. It is an indispensable tool for government agencies, community conservation groups, and environmental organizations seeking to regulate, monitor, and take steps to stop such pollution.

In developing a geographic question related to your own community's potential risk from point-source pollution, keep the following guidelines in mind:

- *Select an appropriate topic.*

  Your study may be triggered by a specific event that has been reported in the news such as a chemical spill, or by the identification of a potential problem such as the presence of toxins in the ground near a former manufacturing plant.

- *Conduct preliminary research to clarify your study goals.*

  It is important to narrow the study's focus to a single issue. Conduct preliminary research into the topic to help develop a hypothesis and clarify the questions you wish to answer. Some questions could include: How many people live within a certain radius of the pollution point? What are the demographic and environmental characteristics of the surrounding area? What nearby natural resources may be affected by the pollution hazard?

- *Define the extent of your study.*

  Some questions can be answered through the study of a single neighborhood (e.g., characteristics of the neighborhood surrounding a gas station from which toxic chemicals have leached into the soil), while others may require the exploration of data from several neighboring communities (e.g., communities within a watershed affected by pollution from an oil storage facility).

- *Form partnerships.*

  Before you begin, talk with local officials, agencies, and groups (Department of Environmental Management, Environmental Protection Agency, watershed protection groups, Audubon Society, Sierra Club, and so on) about developing partnerships for the acquisition of appropriate data, identification of resources for your research, and dissemination of your study's findings. The partnerships you form will positively affect your ability to acquire geographic resources and provide an outlet for reporting your findings.

ACQUIRE GEOGRAPHIC RESOURCES

There are essentially two categories of data that you will need to study point-source pollution issues in your community. First, in order to analyze the areas affected by the pollution hazard, you will need base data about your community such as population characteristics (age distribution, income, density), land use (residential, commercial, industrial, agricultural), physical features (rivers, ponds), and public facilities (library, schools, government offices). Second, you will need data about the point-source pollution sites themselves, including location and any other available relevant information.

Depending on ownership and copyright issues, you may have access to the data you need without charge or you may have to pay a fee to acquire it. Here are some guidelines to assist you in acquiring appropriate data for your study.

Obtaining data about your community

- If your town's planning office uses GIS, they will be the ideal source of data about the characteristics of your community.

- If your town does not use GIS, you may be able to locate appropriate data through your county or state government's office of economic development or regional or statewide planning departments.

- GIS users in your town or region (utility companies, phone company, and so on) may also be willing to provide some of the data you need.

- Your state's Web site may identify available sources of local data.

The table below includes additional suggestions for specific community data.

TYPE OF DATA	SUGGESTED SOURCES
Population characteristics of your town	• Census data from American FactFinder Web site *(factfinder.census.gov)*
Street data for your town	• ESRI Data & Maps Media Kit • U.S. Census Bureau TIGER/Line data from *www.geographynetwork.com*
Land-use data for your town	• Your state's department of environmental management • Your state's department of planning • Aerial photographs of your community • USGS National Land Cover Dataset from *landcover.usgs.gov*
Public facilities data for your town	• Municipal government offices
Bodies of water (rivers, streams, wells, and so on)	• U.S. Census Bureau TIGER/Line data from *www.geographynetwork.com* • USGS National Hydrography Dataset (NHD) from *nhd.usgs.gov* • Your state's department of environmental management
Toxic chemical releases and waste management	• EPA Toxic Release Inventory (TRI) from *www.epa.gov/tri*

Obtaining data about point-source pollution

- Contact local environmental groups—particularly ones that focus on environmental issues related to the subject of your study (watershed protection groups, groups concerned with the health issues related to specific chemicals, groups that focus on hazards associated with specific industries such as power generation, and so on).

- Often you can find data available on the Internet from government sources such as the Environmental Protection Agency or from private environmental organizations such as Lake Ontario Keeper.

EXPLORE GEOGRAPHIC DATA

Technical issues

Whenever possible, ask for data in shapefile format. This is the file format that ArcView can read and display. If the site data that you acquire is not already in this format, you will need to prepare a data table and add it to the ArcView project that displays your community characteristics data. This can be accomplished in two ways:

- Geocoding if you have street addresses for your sites

- Adding an event theme if you have coordinate positions for each site

Preliminary data exploration

It is important to spend some time exploring the data and relationships among the data sets to identify spatial patterns and potential issues for further analysis. Suggestions for preliminary exploration:

- Use the attribute table to explore the nature of the point-source pollution sites data. Depending on the data fields, sort the table by different fields (date closed, classification, and so on) and select records to observe distribution patterns in the project view.

- Use the Legend Editor to symbolize data in a variety of ways:

  - Make copies of the population theme and then use graduated color to symbolize various population characteristics.

  - Use graduated symbols or unique values to symbolize point-source pollution sites according to characteristics in the attribute table such as classification.

- Use the Select By Theme function to identify issues of proximity in relation to the point-source pollution sites theme (within a distance of, completely within, and so on). Identify physical features, landmarks, or land-use areas (such as parks) that are in close proximity to the sites.

- If you are using ArcView 3.2 or higher, use the Create Buffer function to further explore proximity issues. Create buffers of different sizes around point-source pollution sites, landmarks (such as schools), or features (such as rivers).

- Try to identify potential relationships between data sets. For example, are all the point-source pollution sites located in areas with a particular demographic characteristic (age, income, and so on)?

ANALYZE GEOGRAPHIC DATA

In order to analyze your data, determine the preliminary questions that you will need to answer in order to answer the broader, essential question of your study. For example, if your essential question is "Which rivers and neighborhoods are at greatest risk from abandoned landfills in East York?", you might identify and explore the following preliminary questions.

1 Where are the landfills?

   – Add street and landfill location data.

   – Symbolize landfill data as points.

   – Perform visual analysis or use the Identify tool to locate addresses.

2 Which landfills are within 0.25 kilometer of a river or stream?

   – Add landfill and river/stream data.

   – With river/stream data active, use theme-on-theme selection to identify all rivers and streams that are within 0.25 kilometer of a landfill.

   – Identify each of the selected landfills to create a list.

3 Which landfills are located within residential areas?

   – Add land-use data.

   – Query land-use data for all land in the category "residential."

   – Use theme-on-theme selection to find all landfills that are located completely within the selected land use.

   – Identify each of the selected landfills to create a list.

4 Which schools are located within 0.25 kilometer of a landfill site?

   – Add landfill and schools data.

   – Use theme-on-theme selection to identify all schools that are within 0.25 kilometer of a landfill.

   – Identify each of the selected schools to create a list.

5 Where are the A3 landfills compared to the A5 landfills?

   – Add landfill data.

   – Symbolize the landfill sites according to Class of Danger.

   – Observe spatial patterns of the two class designations.

For ArcView 3.1, 3.2, or 3.3 users: With increased functionality in these versions of ArcView software, you are able to take advantage of the Geoprocessing Wizard. You can use this wizard to calculate buffers; perform unions, clips, and intersects; and assign data by location to complete your analysis. Refer to ArcView help for more detailed information. Search for "geoprocessing."

Drawing conclusions

Your research in early stages of the project will guide you in drawing conclusions from your data. For example, the knowledge that landfills of different types differ in the extent (radius) of their impact will shape the conclusions you draw from your GIS analysis. As you develop conclusions and recommendations, you will want to illustrate those conclusions with appropriate maps. The following list provides examples of various maps that could illustrate your conclusions about the questions listed above.

- Location of landfills within the study site
- Landfill sites by risk classification
- Landfill sites by date closed
- Landfill sites within 0.5 kilometer and 0.25 kilometer of schools, hospitals, parks, rivers, streams, ponds
- Schools, hospitals, parks, rivers, streams, ponds within 0.5 kilometer and 0.25 kilometer of a landfill site
- Patterns of landfill distribution within the study area
- Demographic characteristics of residential communities containing abandoned landfills (income level, population density, age structure)
- Landfill sites prioritized by analyzed risk to rivers and streams and surrounding populations
- Recommendations for further environmental impact studies

ACT ON GEOGRAPHIC KNOWLEDGE

The ultimate value of your study is determined by the real changes and actions that it triggers. Those actions will only take place if your conclusions reach the appropriate audience: municipal decision makers, environmental action groups, and the public. The greatest ally of point-source polluters is public ignorance of, and apathy about, the critical issues of the case. Conversely, the most powerful influence against those polluters is a public that is aware and active.

Once you have identified the conclusions and recommendations of your study, you need to present them in an appropriate format to an interested audience or audiences.

Possible action steps

- Prepare a report document, including recommendations suggested by your conclusions, and present it to municipal officials.
- Prepare a multimedia presentation to give in a public forum to raise awareness of the pollution issue addressed in your study. Write a press release to inform local media about the presentation.
- Provide appropriate environmental groups with maps, tables, and graphs that illustrate your study's conclusions so they can be included in educational programs and printed material.

- Invite the local press to write an article about your study. Provide them with maps illustrating your conclusions and recommendations.

- Create a local "watchdog" group to monitor evidence of point-source pollution (water or air quality monitoring, periodic soil sampling, and so forth).

- Contact environmental groups and communities facing similar pollution issues to explore the possibility of expanding your study.

Next step

A point-source pollution study is not an end in itself, but a springboard for further investigation. In addition to drawing conclusions and making recommendations, the study also raises new questions and new avenues for further inquiry. The next step in your own community study will be influenced by responses to your initial efforts and by the specific problems and institutions that will influence the issue's resolution. With each new layer of inquiry and action, you will make a real and enduring contribution to your community and demonstrate the power of one person or group to make a difference.

MODULE 5 ACKNOWLEDGMENTS

Thanks to the World Issues students and their teacher, Rex Taylor, at Crescent School for providing us with this case-study story.

Photographs and supplementary information found in this case study provided by Lake Ontario Keeper.

East York streets, schools, land-use, topographic features, census boundaries, and 1996 population data provided courtesy of DMTI Spatial, Inc. Copyright © 2002.

East York abandoned landfill sites data provided courtesy of Crescent School and Lake Ontario Keeper.

Getting kids to school

It's a simple process to take the bus to school—you wait at the designated location at the right time and the bus arrives. You get on the bus, chat with your friends or finish up some last-minute homework, and the bus takes you to school so you're just in time for your first class. Did you ever stop to think about the planning and coordination behind the scene that makes it possible for school-bus drivers to pick students up and drop them off in the right places at the right times each day? Many geographic questions are involved. Where do the children live? Which ones live close enough to school to walk? Where should the buses stop? Which children should bus number 4 pick up? What is the most efficient route for the bus to take?

Transportation professionals use GIS every day to find the answers to questions like these. They plan routes and schedules for school buses, city buses, delivery trucks, service vehicles, and so on. They take into consideration many factors including the number of buses available, how many people need to ride the bus, and where they need to go. GIS helps them analyze complex information, determine the most efficient routes and schedules, and present this information on a map so people can understand it.

CASE STUDY

Who walks and who takes the bus?
Students at John McCrae Secondary School in Ottawa took on a special project to help out the local elementary school. They used GIS to determine which John Young Elementary students are eligible to ride the bus based on the school's busing guidelines.

EXERCISE

Use buffers to identify eligible school-bus riders
Create buffer zones to reflect school busing guidelines, and then use a variety of analytical tools to select students who are eligible to ride the bus.

ON YOUR OWN

Learn how to establish a partnership with your school or another school to create bus service zones like in the case study. Important data issues, such as data accuracy and privacy issues in using real student information, are included.

Who walks and who takes the bus?

Ottawa, Ontario, Canada

Across North America, schools need to transport students to and from school. In Ottawa, Canada, a group of twelfth graders at John McCrae Secondary School used GIS to help a local elementary school solve its problem of determining which students should walk to school and which should ride the bus. Before the project, the school received requests to ride the bus from parents and students, but they didn't have a way to accurately and systematically measure the distances from students' homes to the school and so couldn't respond.

In Richard Grignon's Geomatics class, students examine the use of technologies such as cartography, surveying, remote sensing, and GIS. The emphasis is on applying what students learn to real-world situations in business, government, or the local community. Each year, students propose a way to use GIS to solve a problem. In spring 2001, a group of geomatics students conducted their authentic project with Judith Toews, principal of John Young Elementary School. Their goal was to produce accurate information about transportation for 552 elementary school students.

IDENTIFYING THE PROBLEM

While working closely with Ms. Toews, the students narrowed the problem of student transportation to two essential questions.

QUESTION	WHAT THEY NEED
What is the road distance to each student's house?	• Student addresses • A way to calculate road distance • Location of school
What is the age of each child?	• Student grades or ages • For each grade or age, the distance at which students are eligible to ride the bus according to school guidelines

Before they began acquiring the necessary data, the students obtained the school's transportation guidelines. This information became the cornerstone from which all analysis was conducted.

GRADE LEVEL	WHEN STUDENTS GET BUSED
Junior and senior kindergarten	Always bused
1–3	Beyond 1.0 km
4–6	Beyond 1.6 km

ACQUIRING DATA

Once the essential questions were identified, the students began to gather necessary data. First, they obtained street data from the City of Ottawa GIS.

Next, they requested student addresses and grade level from John Young Elementary School. This data wasn't as easy to come by. The school was not willing to give the high school students this confidential information because of privacy issues.

Instead, John Young Elementary School and the geomatics students came up with a compromise that would protect the identity of the elementary school children and also allow the high school students to perform their analysis. The school database administrator randomly assigned each elementary school student a student ID number that was different from any number the school previously used. The high school students received data that included student ID, address, and grade. In this way, the high school students received the data they needed, and the school protected the privacy of the individual students by not releasing student names.

John Young Elementary School supplied a student ID, address, and grade for each of its 552 elementary students. With the student information, the geomatics class was ready to begin preparing the data and analyzing it.

Student_id	Address	Grade
1	22 HALKIRK	6
2	54 JAMES LEWIS	6
3	4208 LIMESTONE	6
4	23 NAIRN	6
5	131 PINE RIDGE	6
6	9 SHEPPARD'S	6
7	6539 CRAIGHURST	5
8	118 LEADON	5
9	40 CANBURY CR	4
10	3593 DIAMONDVIEW	4

iyes_students.dbf

PREPARING THE DATA FOR ARCVIEW

Before the geomatics students could determine which elementary students should ride the bus, they needed to map the students' addresses. The geomatics students geocoded, or matched, the addresses in the table to a street data layer. Using detailed street data provided by the City of Ottawa GIS, they geocoded the address data to create a map of student residences.

Students were able to achieve a 95-percent match on the student addresses. The remaining 5 percent of the addresses were for students who live outside the school boundary or who had their own transportation arrangements. Once the geocode was complete, the students began to use the ArcView Network Analyst extension to set up their bus routes from the school to student homes.

The pattern of geocoded student addresses (orange dots) is clustered around John Young Elementary School (symbolized by the star). This map was produced by overlaying an orthophotograph of Ottawa with streets, student addresses, and the location of John Young Elementary School.

FINDING SOLUTIONS

Using Network Analyst, the students created service areas to reflect the school board's three zones of distance to school (0–1 kilometer, 1–1.6 kilometers, and farther than 1.6 kilometers). Each service area mapped the road distance from the school outward.

All student addresses within the service area shown on this map are within a 1-kilometer walk to the school. According to the school board's transportation guidelines, all students except kindergartners within this area are required to walk to school.

Once the three service areas were created, the geomatics students were ready to query the student data to find out how many students were within each service area and how many of those students needed to ride the bus to school.

A query for junior and senior kindergarten students (the only students within 1 kilometer who are required to ride the bus) resulted in 77 students selected. The yellow dots in the map and yellow records in the attribute table represent the kindergarten students who must ride the bus and who live in the 1-km service area.

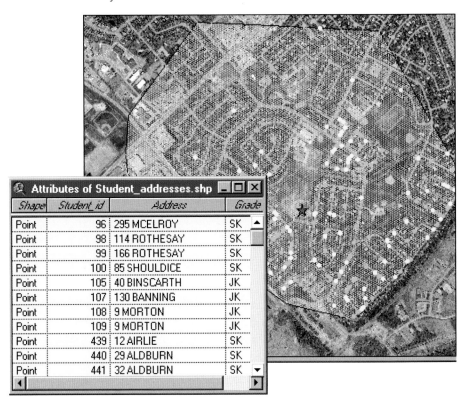

The GIS students queried each of the three service areas and generated a list of student IDs representing students who needed to ride the bus to school.

This summary table shows how many students from each grade needed to be bused. In total, the GIS students calculated that 277 students need bus transportation.

Grade	Count	First_BUS
1	24	bussed
2	29	bussed
3	30	bussed
4	13	bussed
5	22	bussed
6	21	bussed
JK	70	bussed
SK	68	bussed

MAKING THE INFORMATION ACCESSIBLE TO THE ELEMENTARY SCHOOL

After their analysis, the students presented their results to Principal Toews. In preparation, the geomatics students created an ArcExplorer™ project of all their data so she would be able to view the digital maps in the future. In addition, they wrote instructions for her on how to join their student address table with her table that contains student names so that she could generate a list of names of students who needed to be bused. From this list, Ms. Toews would be able to send a letter notifying each student's family whether they will receive bus transportation.

Geomatics students review their analysis results in preparation for their presentation to the school principal.

During the presentation, the geomatics students explained their analysis process and produced a summary table that showed how many students from each grade needed to be bused. In addition, they gave Ms. Toews a list of student ID numbers for those students who need to ride the bus. With this list, she can easily match student ID numbers with her list of student names.

Principal Toews commented, "This project created a positive link between a secondary school and an elementary school. It was extremely beneficial to have quick access to accurate information regarding student distance from home to school."

FUTURE IMPLICATIONS

Since the original presentation, Ms. Toews has used the ArcExplorer project to make empty-seat policy decisions at the school. Quite often, parents will call the school, requesting that their child be allowed to ride the bus, even if they are within walking distance. Ms. Toews uses the ArcExplorer project to look up each student request and make an informed decision about which students would be able to fill the empty bus seats and which would not.

In the future, Mr. Grignon and Ms. Toews plan to have geomatics students map out safe walking routes to school. Also, to help alleviate traffic and parking congestion at the start of each school day, the geomatics students will propose alternate driving routes for private cars.

The geomatics students will continue to work closely with Ms. Toews on improving transportation at the elementary school.

Proud students display their GIS award with the principal of John McCrae Secondary School. The award was presented by the Canada Centre for Remote Sensing in recognition of the students' GIS work on the bus transportation project.

SUMMARY

ASK A GEOGRAPHIC QUESTION	• Where do all the elementary students live? • Where are the service areas for 1-kilometer, 1.6-kilometer, and greater than 1.6-kilometer travel distance from the school? • Which students live in each service area? • Which students need to ride the bus to school because of their age or distance from school?
ACQUIRE GEOGRAPHIC RESOURCES	• Obtain a digital orthophotographic image of the town. • Obtain student information (student ID, grade, address). • Obtain street data for the area around the school.
EXPLORE GEOGRAPHIC DATA	• Geocode student addresses to the street file. • Determine service areas for 1-kilometer, 1.6-kilometer, and greater than 1.6-kilometer travel distance from the school.
ANALYZE GEOGRAPHIC INFORMATION	• Query the data to determine how many students from each service area need to walk to school. • Create a table that summarizes how many students from each grade need to ride the bus. • Prepare a list of students by ID numbers who need to ride the bus.
ACT ON GEOGRAPHIC KNOWLEDGE	• Summarize your results. • Prepare a presentation for the school principal. • Prepare a project so the school principal can explore the data on her own. • Prepare presentations and practice them. • Give presentations.

Use buffers to identify eligible school-bus riders   *Exercise*

School districts typically set busing guidelines according to two factors: the distance a student lives from the school, and the student's grade level. To determine which students live within each of the distance limits stated in the busing guidelines, the students in the case study created distance zones around the school.

In the following exercise you will explore two different methods of measuring distance when creating zones. First you will measure distance "as the crow flies," using a straight line. With this method, you will create a series of circles, or buffers, where the school is at the center and each circle's radius equals a specified distance. Later, you will work with zones created by measuring the length of streets to represent driving distance. The geomatics students created these zones, referred to as service areas, with the ArcView Network Analyst extension.

The ✎ icon indicates questions to be answered. Write your answers on a separate sheet of paper.

Imagine that the principal of John Young Elementary School in Kanata, Ontario, has learned that you enjoy using GIS to solve geographic problems. She has asked you to assist the school in determining which students are eligible to ride the bus to school and which students must walk according to the school's busing guidelines. The guidelines, summarized in the table below, are based on the student's grade level and the distance he or she lives from school.

Busing guidelines

GRADE	ACCEPTABLE DISTANCE
Junior and senior kindergarten (JK and SK)	All bused regardless of distance from school
Grades 1–3	Live 1,000 meters or more from school
Grades 4–6	Live 1,600 meters or more from school

You will provide the principal with the total number of eligible bus riders in each grade and a map showing where the eligible bus riders live. The information will be used to determine the number of buses needed and to plan routes for each bus.

PART 1   EXPLORE STUDENT DATA AND DRAW BUFFERS

ASK

Begin by listing the geographic questions you need to investigate. This list will be a useful reference later when you are ready to analyze the student data.

✎   **1 Write three geographic questions that you could investigate based on the above scenario.**

ACQUIRE

The school and the city have agreed to help you by providing the data you need for your GIS project. You have obtained the following data:

DATA SET DESCRIPTION	DATA FORMAT	FEATURE TYPE	FILE NAME
John Young School: Location of John Young Elementary School	Shapefile	Point	jy_school.shp
Students: Geocoded data for all students of the school (grades JK–6)	Shapefile	Point	students.shp
Roads: Roads for Kanata	Shapefile	Line	kanata_roads.shp
Kanata: City boundary	Shapefile	Polygon	kanata.shp

These files have already been added to an ArcView project, which you'll open now.

2 **Start ArcView. Click the File menu and choose Open Project. Navigate to the exercise data folder (C:\esri\comgeo\module6) and open *bus.apr*.**

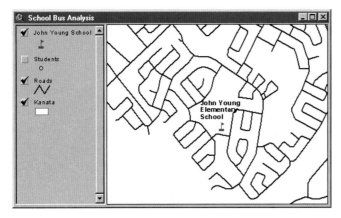

The School Bus Analysis view shows John Young Elementary School and the surrounding neighborhood. You will display and explore the student locations.

EXPLORE

3 **Turn on the Students theme. Click the theme to make it active, then click the Zoom to Active Theme button.**

 3a Where do the students live in relation to the school? For example, do more students live near the school or far from it? Are they evenly spread out in all directions? Describe any patterns you see.

ANALYZE

You are ready to begin your geographic analysis. You know from the busing guidelines given to you by the school principal that you need to consider two factors: *grade* and *distance from school.* You will work with the distance factor first.

In the next steps, you will draw two circles around the school, one with a 1,000-meter radius and the other with a 1,600-meter radius. First you will set the view's distance units so that ArcView reports distance in meters.

4 **From the View menu, choose Properties. Choose meters for both Map Units and Distance Units. Click OK.**

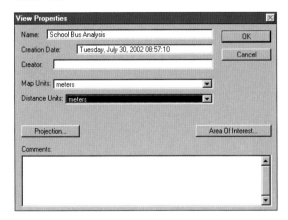

Next you will create a new theme for the 1,000-meter zone.

5 **Turn off the Students theme. From the View menu, choose New Theme. In the New Theme dialog choose Polygon for the feature type and click OK.**

6 **Navigate to the location where you want to save the new shapefile (e.g., *C:\esri\comgeo\module6*). If you are in a classroom environment, ask your instructor where you should save your files for this exercise. Name your new theme 1000m_abc.shp where "abc" represents your initials. Click OK.**

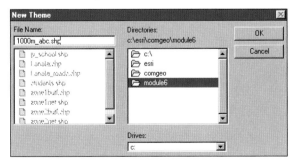

A new theme is added to the table of contents. The dotted line around the theme's check box tells you it can be edited.

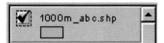

7 Select the Draw Circle tool from the Draw tool pull-down menu. Place the crosshair on the school, then click and drag away from the school. Release the mouse when the radius size shown at the bottom of the ArcView window is close to 1,000 meters. (ArcView may not let you stop on exactly 1,000 meters. You will be able to measure more precisely if you enlarge your view window.) If you need to adjust your circle, click the Pointer tool and move the handles at the edge of the circle, or press the Delete key and start over. You now have a buffer zone around the school with a radius of approximately 1,000 meters.

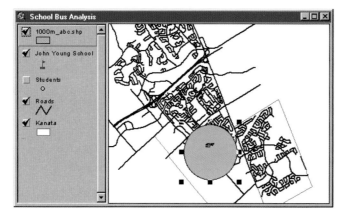

8 Click the Theme menu and choose Stop Editing. Click Yes to save your edits. Your circle may be selected. If it is, click the Clear Selected Features button.

9 Drag the 1000m_abc.shp theme down between the Students and Roads themes in the table of contents.

10 Repeat steps 5–7 to create a shapefile named 1600m_abc.shp containing a buffer zone approximately 1,600 meters around the school.

11 Save your edits.

12 Click Clear Selected features and drag the 1600m theme below the 1000m theme.

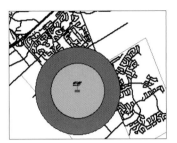

It would be helpful to be able to see the roads and city boundary through the buffers. You will change their solid fill symbols to semitransparent.

13 Open the Legend Editor and Fill Palette for the 1000m_abc.shp theme. Select the symbol in the third row, first column. In the Color Palette, change the foreground color to bright turquoise and the background color to transparent. Click Apply.

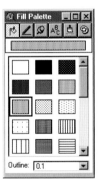
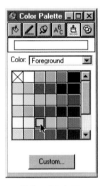
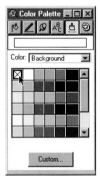

Fill Palette:
Semitransparent

Color Palette:
Foreground to turquoise

Color Palette:
Background to transparent

14 Change the 1,600-meter buffer symbol in the same way. Use transparent for the background color. Use dark green for the foreground color. When you are finished, close the Legend Editor and Symbol Palette.

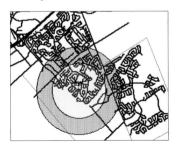

Now you will determine how many John Young Elementary School students live in each zone. To begin, you will find out which students live within zone 1, represented by the area inside the 1,000-meter buffer.

15 Turn on the Students theme and make it active. From the Theme menu, choose Select By Theme. Use the drop-down lists to Select features of active themes that **Are Completely Within** the selected features of **1000m_abc.shp**.

✳ NOTE: Because the buffer theme has only one feature, you don't have to select the feature.

Click New Set.

All the students who live within 1,000 meters of the school are selected.

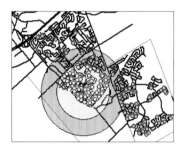

16 Open the Students theme attribute table. Move the Attributes of Students table so that you can see the map and table at the same time. (If necessary, enlarge your ArcView window.) When the attribute table is active, the number of selected records is displayed at the top left of the ArcView window.

✎ 16a How many students live within 1,000 meters of the school? How many students are there in all? Complete a table like the one below. (You will fill in zone 2 and zone 3 information later.)

ZONE	DISTANCE	NUMBER OF STUDENTS
Zone 1	Within 1,000 meters	
Zone 2	Between 1,000 and 1,600 meters	
Zone 3	Beyond 1,600 meters	
	Total students	

Next, you will determine how many students live in zone 2 (beyond 1,000 meters but closer than 1,600 meters). To select these students, you cannot simply select all of the points within the 1,600-meter buffer because it also includes the area within 1,000 meters. Instead, you will switch the current selection to eliminate the zone 1 students before you select by theme with the 1,600-meter buffer.

17 Click the Switch Selection button. The selected set switches to the students living more than 1,000 meters from the school.

18 Make the view active. From the Theme menu, choose Select By Theme. Use the drop-down lists to Select features that **Are Completely Within** the **1600m_abc.shp** theme. This time, click Select From Set.

19 Click the Open Table button to bring the table to the front and display the number of selected records.

19a Use this information to complete the table you started in question 16a. You will need to calculate the number of students who live in zone 3 (beyond 1,600 meters).

20 Close the Attributes of Students table and click the Clear Selected Features button.

This is a good point to stop and save your work.

21 If you are in a classroom environment, ask your instructor how to name your project and where to save it. Make the Project window active and choose Save As from the File menu. Otherwise, click Save. Write down the project name and where it is stored.

PART 2   ANALYZE STUDENT DATA ACCORDING TO DISTANCE BY ROAD

Creating circular buffers as you have just done is a way to analyze distances from the school using basic ArcView drawing and editing tools. The circles indicate distance in a straight line away from the school. But who walks a straight line? In reality, students walking or riding the bus travel along the roads.

Another way to create distance zones is to add up the length of the road segments as you move away from the school. This procedure can be done with the ArcView Network Analyst extension. In the steps below, you will continue your analysis using buffers created in Network Analyst by the geomatics students in the case study.

1 If necessary, start ArcView and open the project you saved in part 1. Turn off the 1000m and 1600m buffer themes and the Students theme.

 2 Click the Add Theme button. Navigate to the exercise data folder *(C:\esri\comgeo\module6)* and add the four themes listed in the table below.

SHAPEFILE	DESCRIPTION
zone1area.shp	Buffer zone 1,000 meters from the school
zone1net.shp	Road network within 1,000 meters from the school
zone2area.shp	Buffer zone 1,600 meters from the school
zone2net.shp	Road network within 1,600 meters of the school

Before continuing with your analysis, you will take a few moments to organize and symbolize the new themes. If you're using ArcView 3.0a, you will need to load the legends for the new themes. If you're using ArcView 3.1 or higher, they were loaded automatically, so skip to step 4.

3 For each of the four themes, open the Legend Editor and click the Load button. Load the appropriate legend file from the exercise data folder. For example, for the Zone1net.shp theme, load zone1net.avl. Remember to apply the changes. Close the Legend Editor when you are finished.

4 Rename the four themes using the following list as a guide.

Zone1area.shp	Zone 1 area
Zone1net.shp	Zone 1 roads
Zone2area.shp	Zone 2 area
Zone2net.shp	Zone 2 roads

5 Reorder the four themes so that your table of contents matches the one shown below.

6 Turn on the Zone 2 roads and Zone 2 area themes. Zoom in to their extent.

7 Any place along the purple roads can be reached by traveling no more than 1,600 meters from the school. The service area was created by connecting the outermost points on the Zone 2 road network. As a result, the service area includes a few streets that slightly exceed the 1,600-meter distance.

8 Turn on the 1600m_abc.shp theme. Pan or zoom the map if necessary.

8a Compare the locations of the 1,600-meter buffer and 1,600-meter service area. Could the method you use to create distance zones affect your analysis results? Why or why not?

9 Turn off the Zone 2 roads and 1600m_abc.shp themes. Turn on the Zone 1 area and Students themes.

You will add a field to the Students theme attribute table where you will record whether or not a student may take the bus.

10 Open the attribute table for the Students theme. From the Table menu, choose Start Editing.

11 From the Edit menu, choose Add Field. In the Field Definition dialog, enter **Bus** for the name, choose String as the type, and enter **5** for the width. Click OK.

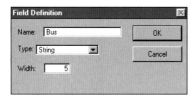

12 Scroll to the end of the table to see the Bus field. Notice that it is blank for all of the students.

You will enter No in the Bus field for all students. Then, as you proceed with your analysis, you will change the attribute to Yes only for those students in each zone who may ride the bus.

13 Click the Select All button to select all the students.

14 In the table, make sure the Bus field is active, then click the Calculate button. In the Field Calculator text box type **"No"**. Be sure to include the quotation marks, as in the following graphic, and then click OK.

The word No is added to each row in the table.

You will analyze the student data geographically, one zone at a time. You'll begin with Zone 1.

 14a Refer back to the Busing Guidelines at the beginning of the exercise. Which grades are correct in the table below?

Zone 1	Distance from school: 0–1,000 meters							
Grades that ride the bus	JK	SK	1	2	3	4	5	6

15 Make the View window active. Make sure that Students is the active theme. From the Theme menu, choose Select By Theme. Use the drop-down lists to Select features of active themes that **Are Completely Within** the selected features of **Zone 1 area**. Click New Set.

All students who live within zone 1 are selected, but only those who are in junior or senior kindergarten may ride the bus. You will perform a query to modify the selected set of students.

16 Click the Query Builder button. Do the following actions to build the query expression to select the kindergarten students:

- Double-click [Grade] in the Fields list

- Click once on the "=" button

- Double-click "JK" in the Values list

- Click the "or" button

- Double-click [Grade]

- Click "="

- Double-click "SK"

Make sure your expression matches the one in the graphic below, then click Select From Set.

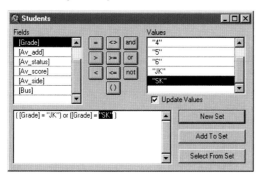

17 Close the query builder window. Use the map and table to answer the following questions.

17a What happened to the selected set of students on your map?

17b How many kindergartners live in zone 1?

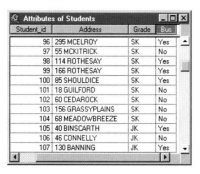

18 Make the table window active, and make sure the Bus field is active, too. Click the Calculate button. Type **"Yes"** in the text box. Click OK.

The word **No** is replaced with **Yes** for each student in the selected set.

18a What are the correct grades for zone 2?

Zone 2	Distance from school: 1,000–1,600 meters							
Grades that ride the bus	JK	SK	1	2	3	4	5	6

To select the students in zone 2, you will use the same procedure you used in part 1 of the exercise. First you will select the zone 1 students, switch the selection, then eliminate the zone 3 students.

✳ **NOTE:** It is possible to select the zone 2 students using different combinations of steps, but the method you will use takes fewer steps than other methods. You might think it would be more logical to select zone 2 students, then select and remove zone 1 students from the set. This method is not possible because the Select By Theme dialog does not offer a Remove From Set button.

19 Make the view window active. Make sure the Students theme is active. From the Theme menu, click Select By Theme. Select features of active themes that **Are Completely Within** the selected features of **Zone 1 area**. Click New Set.

20 Make the attribute table window active, and click the Switch Selection button. Now the zone 2 and 3 students are selected.

21 Make the view window active again. From the Theme menu, choose Select by Theme. Choose **Are Completely Within** and **Zone 2 area** from the drop-down lists. Click Select From Set.

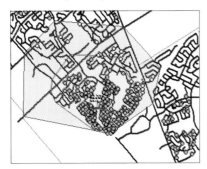

Your selected set now includes only the students in zone 2.

21a Refer to question 18a and complete the following expression that will select the zone 2 students who may ride the bus. Choose from among the **<, >,** or **=** symbols.

([Grade] _____ "3") or ([Grade] _____ "JK")

✳ **HINT:** Numerals and letters in a string field are sorted so that 0–9 come before A–Z. Even though you know "JK" and "SK" come before grade 1 because students take kindergarten before grade 1, the computer lists these codes after grade 6 in your student table. You must take this into account when you build your query expression.

22 Click the Query Builder button. Use the tools in the dialog to build your query expression. Make sure to include parentheses at the beginning and end of the expression. Click Select From Set.

✳ **NOTE:** If you make a mistake and get an error, delete the expression from the text box and try again.

22a How many students in zone 2 are eligible to ride the bus?

23 Make the table active and click the Calculate button. Calculate Bus = "Yes" for these students.

23a What are the correct grades for zone 3?

Zone 3	Distance from school: Beyond 1,600 meters							
Grades that ride the bus		JK	SK	1	2	3	4	5	6

24 Zoom to the extent of the Students theme.

25 Use theme-on-theme selection to select all the students who live in zone 3.

✳ **HINT:** Use the zone 2 buffer, then switch the selection. For help, refer to steps 18 and 19.

25a How many students in zone 3 are eligible to ride the bus?

26 In the attribute table, calculate Bus = "Yes" for these students.

27 From the Table menu, choose Stop Editing. Click Yes to save your edits. Clear the selected features.

ACT

Next you will create a table for the John Young Elementary School principal that summarizes the information by grade level. First you will determine the total number of students in each grade. Then you will determine how many of them may ride the bus.

28 Make Grade the active field in the table. Click the Summarize tool. Accept the defaults in the Summary Table Definition dialog and click OK. The summary table lists the total number of students (Count) in each grade.

28a Copy the data into the middle column in a table like the one below.

Eligible bus riders by grade

GRADE	TOTAL NUMBER OF STUDENTS	NUMBER OF STUDENTS ELIGIBLE TO RIDE THE BUS
JK		
SK		
1		
2		
3		
4		
5		
6		

29 When you are finished, close the summary table.

30 Click the Query Builder button. Build a query where Bus = Yes to select all the students who may ride the bus. Click New Set. Close the Query Builder window.

31 Click the Summary button, and then click OK to summarize the selected records.

31a Record the data in the last column of your report table (step 28a).

32 Close the summary table.

33 You will provide the principal with an estimate of the number of buses that will be needed. One bus can transport fifty to sixty students.

33a How many eligible bus riders are there in all?

33b How many buses will be needed to transport them?

In addition to the numeric summary, you want to give the principal a map showing where eligible bus riders live. To do this you will make a new shapefile containing bus riders only.

34 Make the View window active. From the Theme menu, choose Convert to Shapefile. Navigate to the folder where you are saving your work for this exercise, and name the file **bus_abc.shp** where "abc" represents your initials. Click Yes to add the theme to the view.

35 Turn off the Students theme and turn on Bus_abc.shp. Change the theme name to **Eligible bus riders**. If desired, change the symbol color.

36 Create and print a layout for the principal illustrating the three zones, the school, and the eligible bus riders.

37 Save your project.

SUMMARY

In this exercise, you:
- Drew buffers according to specified distances
- Observed differences between buffers based on a radius and service areas based on a network
- Analyzed the student data by zone and grade level to determine who can ride the bus
- Added bus information to a new field in the student attribute table

ASK A GEOGRAPHIC QUESTION

The case study and exercise revolve around the idea of transportation planning. Transportation planning includes finding out who needs transportation services, where these people are located, where they need to go, and when. A related concept is transportation routing—in simple terms, the planning of how to get from point A to point B. Every day we encounter the results of transportation planning, yet we hardly think twice about the work that has been done to make it happen. For every school bus or city bus we see on the road, hours of planning and processing of spatial data are involved in getting that bus to and from its destinations.

Buses aren't the only form of transportation to which planning can be applied. Airlines, trains, delivery trucks, courier services, and many others all need to plan travel routes and schedule stops along the way. GIS maps and tools are useful for finding the most efficient routes and making deliveries or destinations on time.

The following suggestions will help you pose and select geographic questions to investigate.

- Learn the transportation guidelines for your school or school district. How does the school match the guidelines with the student locations? How do parents find out how far they live from the school and whether their children are eligible to take the bus?

- Who would be a logical partner to work with? In the case of school buses, obvious partners include the local school administration, school board, or school database administrator.

- Keep in mind the scope of your geographic question. If you live in a large school district, it could be challenging and time consuming to plan the bus routes to serve a student population in the thousands. Start off small, as in the case study, with one school or a small service area. Check with your school district's administrative offices to find out whether they are already using GIS for transportation planning. If so, they could be an excellent resource and partner.

- Brainstorm various transportation systems within your local community.

- Identify local businesses or municipal departments that work with routing on a daily basis, such as a local package delivery or messaging service, solid waste collection, or water meter readers. Learn what challenges they face on a daily basis and what tools they use to solve them. For example, how do they know where a particular address is located? What is more important to their activity, the number of stops they visit each day, or the order of the stops? Does each driver have a regular route that is the same day after day, or does a driver have completely different stops to make each day?

What transportation issues are important at your school? You can use GIS to map and understand:

➤ Projecting the number of students needing transportation for the next five years

➤ Best locations for bus stops

➤ Assigning students to bus stops

➤ Safe routes for students walking to school

➤ Locations where crossing guards should be posted

➤ Possible walking buddy lists for students who walk

➤ Possible ride-sharing lists for students who carpool

ACQUIRE GEOGRAPHIC RESOURCES

Here are some types of data you may need for your transportation analysis and possible sources of such data.

TYPE OF DATA	POSSIBLE DATA SOURCES
Student addresses and information	• Local school (see "Tips" below for important notes)
Street data for geocoding	• Your community's GIS or city planning department • Your state or county Web site • GIS users in your community (e.g., utility companies, phone companies, etc.) • Data sites on the Internet such as U.S. Census TIGER/Line data from *www.geographynetwork.com* (see the Community Geography Web site *www.esri.com/communitygeography* for additional sources)
Other geographic features	• Your community's GIS or city planning department • State natural resource information office (often available on the Web) • Regional council of governments Web site • U.S. Geological Survey (USGS) for topographic maps and aerial photographs on *www.terraserver.microsoft.com* and other data via *www.earthexplorer.usgs.gov* • USGS and other basemapping data via *www.gisdatadepot.com*

Tips

• Student data is confidential information and should be handled with care. In the case study the students working on the project asked only for the following fields in the data: student ID, address, and grade level. The student ID assigned to this data was randomly assigned and unique from the traditional ID number given by the school. The special student ID was used so that only the principal could identify the individual students, and student privacy was guaranteed.

• Because student data is highly confidential, it may take several weeks to obtain permission to use it. Keep this in mind when planning a time line for your project.

• Request that the data to be given in a tab- or comma-delimited text file so that it can be easily imported into ArcView. For instructions on how to import a table into ArcView and geocode address data, see the module 2 exercise.

• Make sure the street data is compatible with the address data you receive from the school. This will make the process of geocoding the information much easier. (The address table should have the entire address in a single field, while the streets data table should have several fields containing street attributes.)

• Be sure your street data has the required characteristics. For example, to geocode the student locations, you need streets with address ranges. (Refer to module 2 for more tips on geocoding.) If you plan to use the ArcView Network Analyst extension, the street data needs to have proper topology (i.e., streets need to be "connected" in order to measure distance along a path from the school to a student location). If you plan to do more advanced analysis, such as including a travel time factor, you may need an attribute containing an average speed or maximum speed limit.

- Research (using the library or the Internet) for resources that provide a general understanding of the issues involved in transportation routing or planning, any conventional methodologies or approaches that may relate to your project, or ways GIS is applied to transportation-routing problems.

EXPLORE GEOGRAPHIC DATA

- Review each theme, including the theme's attribute table. Does the theme include the fields you requested? Do you have all the data you'll need for your intended analysis? For example, if you must consider the students' grade level, be sure you have that field in your table.

- Create a map showing the student locations, school, streets, and other pertinent themes such as city boundaries, school district boundaries, or parks. Look at the map and observe general patterns as to where students live in relation to the school.

- Draw (manually or automatically) a series of circles or buffers around the school at key distances. These could be the distances specified in the school's busing guidelines, or simply half-mile or half-kilometer increments. Visually analyze the student locations and the street network at the various distances.

ANALYZE GEOGRAPHIC DATA

- Plan the procedure you will use to select the student groups specified in the busing guidelines. For example, will you start with an attribute query to select students in a certain grade range, and then do a spatial query to narrow the selection to a particular zone? Or do you need to do the steps the other way around? Will you start with the younger grades or the older grades? The exercise illustrates an efficient way to do the analysis, but it is not the only way. Depending on the structure of your school's busing guidelines, you may need to modify the procedure for selecting eligible students.

- Decide ahead of time what field or fields you will add to the student table, and decide on the words or codes you will use for the attributes. In the exercise, for example, the field "Bus" was added, and the words "Yes" and "No" were used to indicate whether a student could ride the bus. Alternatively, you may decide to use codes like "1" for bus riders, "2" for walkers, and "3" for train riders.

The students in the case study took the steps below. Use this list as a guide for analyzing your own geographic questions.

1 Where do students live in relation to school?

– Geocode the student address data.

– Create transportation zones reflecting the busing guidelines.

2 Who is eligible to ride the bus?

– Systematically select groups of students by attribute (such as grade) and zone. For each selected set, calculate the appropriate value in a new attribute field.

– Summarize the number of students in each zone who meet the requirements for riding or walking.

– Identify total number of bus riders. Estimate how many buses are needed.

– Create maps showing where the eligible bus riders live for bus route planning.

– Create maps showing where walkers live for planning crossing-guard locations and recommended walking routes.

Tips for creating zones

In the exercise, buffers defining distance from the school are created manually. There are a number of possible ways to create buffers or zones depending on the software you are using and the needs of your analysis.

• You can create new shapefiles and draw your own buffers or zones. Besides being the simplest method, this method is also useful if transportation guidelines use parameters other than distance, for example, arterial streets or subdivisions.

• If you are using ArcView 3.0a, you can automatically create buffers at a specified distance using the sample script extension .avx that comes with the software. A similar tool is used in the module 8 exercise.

• If you are using ArcView 3.1–3.3, you can create buffers automatically using the Geoprocessing Wizard. This tool will automatically create distance buffers around a feature using a wizard that will walk you through the process step by step.

The students in the case study used Network Analyst to create their zones rather than drawing simple distance buffers. Network Analyst creates zones, referred to as service areas, from travel distance along a street network.

ArcView extension useful for transportation analysis

SOFTWARE	WHAT YOU CAN DO WITH IT	WHERE TO GET IT
ArcView Network Analyst extension	Creates service areas based on a specified distance or travel time on a network (e.g., streets). Finds the shortest route between two points on a network (e.g., a school and a student's home).	Available through *www.esri.com* or your local ESRI Business Partner. Educational pricing is available. PC only.

ACT ON GEOGRAPHIC KNOWLEDGE

Once you have developed your solution, you must present your findings to your partner (for instance, the school administration). The students in the geomatics course created a portable project in ArcExplorer for the principal to use (see table below).

Possible action steps

- Prepare a multimedia presentation illustrating the solutions and demonstrating findings in ArcView.

- Inform parents and students about local school transportation issues.

- Make a case as to why the school district should add or remove buses from their fleet.

- Create a portable project for use in ArcExplorer or ArcView, or publish maps for viewing in ArcReader™ (see table below).

- Create a printed map or interactive map or Web site for parents or the public (e.g., real estate agents or potential home buyers) to determine in which school transportation zone a particular address is located.

ArcView software useful for distributing maps

SOFTWARE	WHAT YOU CAN DO WITH IT	WHERE TO GET IT
ArcExplorer	Provides basic GIS tools for displaying and exploring maps based on shapefiles and other GIS data.	Free. Available through *www.esri.com.* PC only.
ArcReader and ArcGIS® Publisher	Allows anyone to view, explore, and print published map files (PMFs). PMFs must be created with ArcGIS Publisher.	ArcReader. Available through *www.esri.com.*

NEXT STEPS

Determine how your analysis has helped the school. What future projects can you and your team work on based on your initial findings? The students in the geomatics class found their plan to be effective. They plan to continue working with the elementary school to look at other transportation issues. The next project will be to do a detailed study of students who walk to school to determine safer routes, where crosswalks should be improved, and where crossing guards should be located. Another issue they may explore is how to manage traffic flow for cars dropping off and picking up students at school.

MODULE 6 ACKNOWLEDGMENTS

Thanks to Richard Grignon and his geomatics students at John McCrae Secondary School for providing us with this case study.

Orthophotograph images are used with permission from National Capital Commission, copyright © 2001.

Student information data and ArcView Network Analyst service area data is provided by Ottawa Carleton District School Board and used with permission. All student data has been modified to provide a simulation database ensuring confidentiality. Information provided in this document does not correspond to actual student information.

Street data, municipal boundaries, and school location data is provided by City of Ottawa, copyright © 2000 and used with permission.

Protecting the community forest

What do New York's Central Park, a New England town, and a midwestern Main Street have in common? At each location, it is trees—the community forest—that give character to the human-built environment and create a unique sense of place. Across the country, community forests are threatened by the phenomenon of urban sprawl and the dangers of unplanned growth. Through community action today we can protect and preserve this vital resource for our future.

CASE STUDY

One, two, tree—Taking a tree inventory

In Barrington, Rhode Island, middle school students wanted to help their town develop a community forestry plan by initiating a community tree inventory. They carefully collected and recorded data for trees on their school property, used GIS to map and analyze their data, and presented their data and conclusions to their town for inclusion in its own growing GIS database.

EXERCISE

Map and query a tree inventory to locate hazardous trees

You will use GIS to explore and analyze a tree inventory. After digitizing trees and other features from an aerial photograph, you will analyze the data to identify current patterns and potential problems in an existing tree population.

ON YOUR OWN

Explore the tasks and resources needed to conduct a tree inventory in your own community and to use GIS to analyze that inventory. By outlining strategies and suggestions for data collection, data analysis, and the presentation of conclusions, the section provides guidelines for community groups wishing to implement a similar project.

One, two, tree—Taking a tree inventory

Barrington, Rhode Island

Recent changes in state planning guidelines require every Rhode Island town to include an urban forestry component in its comprehensive plan. The eighth-grade students at Barrington Middle School helped their town meet these new regulations by taking a tree inventory of their 29-acre school campus.

A tree inventory is an essential prerequisite for community forestry planning. By identifying the location, species, and condition of the existing tree population, the inventory allows town planners to anticipate future arboreal needs and potential problems.

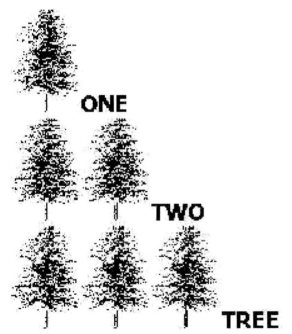

ONE

TWO

TREE

Project One, Two, Tree is funded by a $25,000 grant from the National Geographic Society Education Foundation and by a donation of more than $20,000 in ArcView 3.x software from ESRI. Other One, Two, Tree partners include American Forests, The Rhode Island Tree Council, Rhode Island Department of Statewide Planning, Brown University Urban Studies Program, University of Rhode Island Environmental Data Center and Cooperative Extension, New England Gas Company, Rhode Island Department of Environmental Management Division of Forest Environment, Grow Smart Rhode Island, Moonbeam Publishing, and the Barrington Public Schools.

Under the guidance of geography teacher Lyn Malone and science teacher Jim Kaczynski, eighty-five Barrington Middle School eighth graders have identified, measured, and recorded data on 217 trees on their middle school property. The One, Two, Tree inventory is an interdisciplinary project that merges geography, natural science, and environmental studies with GIS to address a significant community issue—the protection and preservation of community forests.

COLLECTING THE RIGHT DATA

Attributes of Trees

Shape	Area	Perimeter	Tree_id	Pub_pr	Species	Dia_cls	Diameter	Ht_cls	Rd_h	Health	Cnf_pow	Cnf_eon	Cnf_st	Cnf_side	Cnf_tree
Polygon	463.19	76.29	2	B	WA	2	12	3		2	N	N		Y	Y
Polygon	736.12	96.18	3	B	WA	3	31	3		3	N	N		Y	Y
Polygon	2026.02	159.56	4	B	RO	3	50	3		4	N	N		Y	Y
Polygon	328.42	64.24	5	B	GM	2	11	2		3	N	N		N	Y
Polygon	739.51	96.40	6	B	GM	2	15	2		3	N	N		N	Y
Polygon	2414.46	174.19	7	B	RO	3	56	3		3	N	N		N	N
Polygon	452.99	75.45	8	B	WA	2	13	3		2	N	N		N	N
Polygon	2835.74	188.77	9	B	BOK	3	61	3		4	N	N		N	Y
Polygon	263.87	57.58	10	B	CH	1	9	3		4	N	N		N	Y
Polygon	1142.68	119.83	11	B	WA	3	39	3		3	N	N		N	Y
Polygon	902.59	106.50	12	B	BOK	2	17	3		3	N	N		N	Y
Polygon	697.70	93.64	13	B	BOK	2	16	3		3	N	N		N	Y
Polygon	1082.78	116.65	14	B	SYC	3	21	3		3	N	N		N	N
Polygon	1210.21	123.32	16	B	RM	3	38	2		2	N	N		Y	Y
Polygon	229.02	53.65	17	B	GM	2	13	3		2	N	N		N	Y
Polygon	825.13	101.83	18	B	RM	3	32	3		2	N	N		N	Y
Polygon	538.35	82.25	19	B	RM	2	13	2		2	N	N		N	N

Barrington Middle School eighth graders collected all the data displayed in this table. From left to right, the field names represent the following data: data shape, area of tree, perimeter of tree, tree ID number, whether the tree is publicly or privately owned, species type displayed with a code, diameter class of tree where the numbers 1, 2, and 3 represent different diameter ranges, actual diameter, height class, reduced height of tree (if tree was "topped" off), health of tree determined by response to specific questions, and whether the individual tree is or has the potential to be in conflict with power lines, the school building, streets, sidewalks, or other trees. In total, the students collected data on 217 trees.

When designing the project, Ms. Malone consulted with the Barrington Public Works Department to identify the most useful data for tree maintenance. She and Mr. Kaczynski also interviewed foresters and arborists experienced in conducting street tree inventories to identify database content and characteristics that would be useful for community forestry planners. Once they compiled a list of necessary data, they began preparing their students for the task ahead.

Is it a black oak or a white oak? Students consult their tree identification guide to determine the correct tree species.

180

In order to complete this project, students needed to learn how to collect tree inventory data, import it into ArcView, and analyze it. At the kickoff event in September, students were treated with presentations from professional foresters and GIS analysts. These inspirational talks helped students understand how trees are valuable to all communities. They reduce water runoff, soil erosion, and floods; purify water and air; are vital to wildlife habitat; provide recreational opportunities for humans; help counter the greenhouse effect by retaining atmospheric carbon; enhance property values; reduce energy costs by providing shade to buildings; and buffer noise. The professionals who spoke to the students left them with a sense of purpose—that this tree inventory project was important and that their role in it was vital.

After researching the value of trees to communities, students summarized their benefits in the informative posters "Barrington Counts on Trees."

For a week in October, eighty-five students inventoried the 29-acre school campus. Each four-person team began by reporting to their assigned area with materials and data sheets in hand. They picked up leaves, consulted tree identification guides, and debated the species of each tree in their area. Once identified, students used a clinometer to calculate the height of each tree and used a measuring tape for tree diameter and crown radius. On their data sheets, students marked the condition of the tree as excellent, good, fair, poor, dying, or dead. They also observed and noted whether each tree was in conflict with a utility line, a building, the street, the sidewalk, or another tree. Trees that were in poor health and in conflict were later classified as hazardous and recommended for removal.

While measuring a distance of 100 feet from the base of a tree, the students use a clinometer to measure the angle to the top of the tree. The height of the tree is calculated by plugging the angle into an algebraic equation.

Data collection materials include (from top left clockwise) a clipboard for holding data sheets, a clinometer to measure tree height, a tree species identification guide, a small forester's tape used for determining DBH (diameter at breast height), a group number card for each group member, a long measuring tape used in conjunction with the clinometer, and a tree identification tag used to mark each tree.

ANALYZING THE DATA IN ARCVIEW

Students created a basemap for their project by digitizing the school building and trees on an aerial photograph of the school property. As they digitized the trees they had inventoried, students created an attribute table from data collected in the field that linked each tree's characteristics to the polygon on the map. With a completed attribute table, they were able to display and map the tree data with ArcView 3.x. Each student group was responsible for analyzing the data according to four categories: species diversity, tree age, tree condition, and existing or potential problems.

Students created graphs that display tree species diversity. The blue bar indicates that red maple trees are the most prevalent species on the Barrington Middle School campus. The green bar on the left shows that black oak trees are the second-most prevalent.

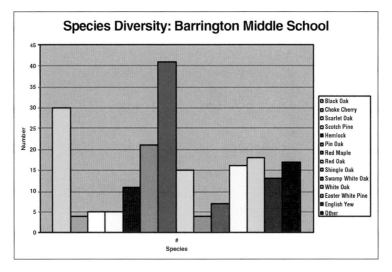

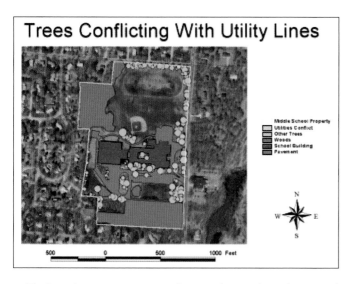

Trees Conflicting With Utility Lines

Middle School Property
Utilities Conflict
Other Trees
Woods
School Building
Pavement

This map was prepared by digitizing the school building and trees on the orthophotograph of the school property. Trees in conflict with utility lines were assigned a different color from other trees to create this thematic map.

Their assignment was to write an observation about each of the four categories, create a map and a graph to illustrate that observation, and create a table displaying the data used to make the maps and graphs. To create many of the maps, students needed to learn how to symbolize data, perform complex queries, sort the data, rename themes, and create layouts. In all, they spent over a week preparing their maps and graphs.

In creating their project, the Barrington students used an ArcView extension called CITYgreen that was developed by American Forests (see more about CITYgreen in the "on your own" section of this module). This extension significantly expanded the ability of ArcView to analyze a tree inventory, but the analysis described in this case study and exercise do not require the extension.

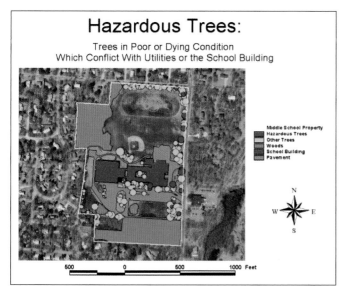

Hazardous Trees:

Trees in Poor or Dying Condition
Which Conflict With Utilities or the School Building

Middle School Property
Hazardous Trees
Other Trees
Woods
School Building
Pavement

Students queried the data to select all trees in poor or dying condition and that conflicted with utility lines or the school building. They converted the selected trees to a new theme named "Hazardous Trees." With their symbol color changed to purple, the hazardous trees were easily visible on the map.

Each group presented a "Tree Inventory Summary Report" to the class that summarized their observations and listed recommended actions for the town. These observations included:

- There are 26 different tree species on the school property, but two-thirds of the trees are either oaks (98) or maples (48). The dominant species, representing 19 percent of the total tree population, is red maple.

- Almost half of the trees (48 percent) are fully mature (having a diameter of 20 inches or greater). Sixty-five percent of the property's 98 oaks are mature, and 58 percent of the property's 48 maples are mature. Young trees represent 22 percent of the total tree population.

- Although the majority of trees on the school property are in fair or good condition, 33 percent of them are dead or dying. Almost half (48 percent) of the property's maples are categorized this way, compared to only 16 percent of the oak trees.

- A number of trees on the school property represent potential hazards, because they conflict with overhead utilities, the school building, sidewalks, and driveways. Of the trees on the school property, 33 conflict with existing utility lines.

PRESENTING RESULTS AND RECOMMENDATIONS

With upcoming presentations to the Barrington Town Council and to the Barrington School Committee, the students needed to summarize their results in a visually effective and informative way. They used their colorful maps and graphs to illustrate the location of the trees on the school campus. In the end, they made a list of recommendations and twelve students presented them to the town council. Their recommendations were:

- Remove trees that are dead or dying—especially those close to the school building or utility lines.

- Inspect and prune trees that conflict with existing utility lines. Those in poor condition may need to be removed.

- Examine trees listed in poor condition to see if remedial intervention can improve their health.

- Plant new trees both to replace those removed and to assure the survival of the property's tree canopy when additional trees are removed in the future.

With a new database of tree information and a list of recommendations, the Barrington Public Works Department is now one step closer to developing a program of tree maintenance and replanting in the town. In 2001–2002, students will conduct a tree inventory of Barrington High School and will add it to the town's growing database.

Proud students and teachers are all smiles after their successful presentation to the Barrington Town Council.

NEXT STEPS

Teachers Lyn Malone and Jim Kaczynski are expanding their One, Two, Tree project throughout Rhode Island and the United States. During the summer of 2001, Ms. Malone and Mr. Kaczynski held a One, Two, Tree Teacher's Institute for Rhode Island teachers interested in completing a tree inventory at their school. The weeklong institute gave teachers practice collecting tree data, an introduction to GIS analysis using ArcView 3.x, and time to review the project lesson plans.

As a result of this institute, five hundred students in ten communities throughout Rhode Island are collecting tree inventories in their community during the 2001–2002 school year.

This is just the beginning. Imagine the benefits if students across the country had the opportunity to do this work in their community. Students in Rhode Island have gained a stronger sense of place and commitment to their community; they became problem-solvers using science, geography, and computer skills; each community is benefiting significantly from the growing database the students are providing; and schools and communities are forming partnerships that benefit all parties. Get involved and make a difference in your community, too!

Teachers at the One, Two, Tree Teacher's Institute review lesson plans for the project and practice tree data collection techniques.

YEAR	NUMBER OF STUDENTS	NUMBER OF COMMUNITIES
2000–2001	85	1 (Barrington Middle School)
2001–2002	500	10 (throughout Rhode Island)
2002–2003	5,000	50 (at the national level)

SUMMARY

ASK A GEOGRAPHIC QUESTION	• Where are the trees on the school property? • Where are different species of trees? • Where are the mature trees? The young trees? • Where are the trees in good, fair, and poor condition? • Where are the trees that are in conflict with utility lines, sidewalks, the school building, or driveways? • Where are the trees that are hazardous?
ACQUIRE GEOGRAPHIC RESOURCES	• Obtain a digital orthophotographic image of the town. • Determine the protocol for collecting tree data. • Collect the following data for each tree: identity number, species, area, perimeter, ownership (public, private, or unknown), diameter, height, health, condition, and conflicts with utility lines, the school building, sidewalks, or driveways. • Record data in a table that can be imported into ArcView.
EXPLORE GEOGRAPHIC DATA	• Digitize the school building and individual trees. • Thematically map species, tree age, tree condition, and existing problems.
ANALYZE GEOGRAPHIC INFORMATION	• Create graphs of species diversity, tree age, and tree condition. • Query the data to map where all mature oak and maple trees are. • Query the data to map where all hazardous trees are (in poor or dying condition and in conflict with utility lines).
ACT ON GEOGRAPHIC KNOWLEDGE	• Summarize your results. • Prepare a list of recommendations for the town council and school committee. • Prepare presentations and practice them. • Give presentations. • Deliver the tree database to the town council.

Map and query a tree inventory to locate hazardous trees

Exercise

To do a tree inventory, data about the trees is first collected "in the field." To be able to analyze that data using GIS, the next step is to build a basemap by digitizing from an aerial photo of the area. Then each tree in the inventory is digitized, including adding descriptive data (also called attribute data) collected during the field study. The map and the attribute data are used to identify and analyze characteristics and potential problems within the existing tree population.

The ✏ icon indicates questions to be answered. Write your answers on a separate sheet of paper.

ASK

Overhead utility lines stretch through the beautiful oaks and maples that surround Barrington Middle School and line nearby streets. The town's Public Works Department has asked you to identify trees that represent a potential hazard: specifically, they are concerned with trees that are close to these utility lines or the school building and are in poor health. Knowing which trees present the greatest risk will guide the department in planning essential tree maintenance. In this exercise you'll create a basemap for a tree inventory, analyze collected tree data to identify hazardous trees, and prepare a map to show their location.

PART 1   CREATING A BASEMAP AND DATABASE

1   Double-click the ArcView icon on your computer's desktop to start ArcView. If you don't have an icon on your desktop, click ArcView from the Start menu.

2   Navigate to the module 7 exercise folder *(C:\esri\comgeo\module7)* and open *tree.apr*.

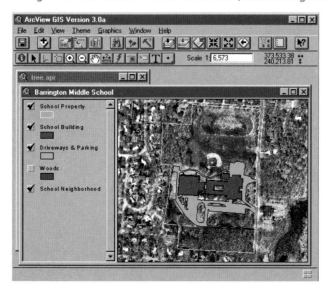

The Barrington Middle School view displays an aerial photograph of the school and its surrounding neighborhood. The School Building and Driveways & Parking themes were created by digitizing, or tracing, from the photograph.

3 **Turn the School Building and Driveways & Parking themes on and off to see the relationship between the visible shapes and the photograph beneath.**

4 **Zoom to the area enclosed in the school's semicircular drive on the east side of the map.**

4a Describe what you see in the area enclosed by the semicircular drive.

5 **Use the Pan tool to look at the area in the southeast corner of the school property.**

5a How would you describe the area between the driveway and the lower property line?

The trees that you see in the southern portion of the image are part of a heavily wooded area. The trees within the semicircular drive are called "street trees." This tree inventory focuses on street trees and free-standing cultivated trees on the school campus.

6 **Make School Property the active theme and use the Zoom to Active Theme button to return to a view of the entire school property. Turn on Woods.**

This theme shows wooded areas on the middle school property that were not included in the school's tree inventory. In the next steps, you will digitize or draw street trees on the school basemap and enter appropriate data about each tree.

7   Zoom to the area enclosed by the semicircular driveway in front of the school. Locate the cross-walk on the street and zoom to the area just north of the crosswalk.

8   From the View menu, click New Theme. For the Feature type, select Polygon. Click OK. Name your new theme **New_tree** and save to the default location or another preferred location on your computer.

New_tree.shp now appears in the table of contents. Notice that this theme's check box has a dotted line around it. This means that the theme is in editing mode and can be modified. Next you will change the legend for New_tree.shp and then you will digitize a tree based on its measured crown radius.

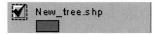

9   Open the Legend Editor for New_tree.shp. Change the symbol color to medium green and click Apply. Close the Legend Editor and the Color Palette.

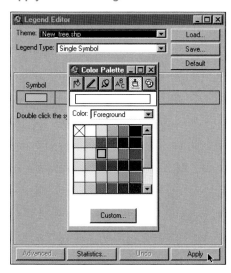

10 From the drawing tools drop-down list, click the Draw Circle tool.

ACQUIRE

11 To digitize the first tree, place your cursor on the base of the first tree that is north of the cross-walk. Hold down the left mouse button and drag a circle with a 21-foot radius from that base (this is the tree's measured crown radius).

⁛ **HINT:** You can see the radius of the circle in the status bar beneath the view. Don't worry if you are off by half a foot or less.

Release the mouse button and you see the tree you have just drawn. What if your circle isn't the correct size? If you create an incorrectly sized circle, press the Delete key and start over.

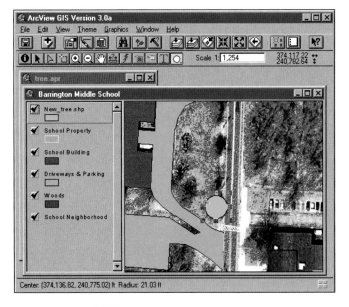

12 Open the New_tree.shp attribute table.

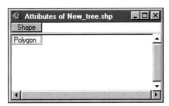

This table currently has one field named Shape. If you are using ArcView 3.2 or ArcView 3.3, another field named ID is automatically created. In the next steps, you will create four new fields and populate them with data about that tree.

13 From the Edit menu, click Add Field. Name the new field **Tree_id**. Give it the following characteristics and click OK.

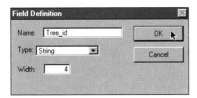

In addition to tree identity, you will create fields for tree species, diameter, and height class.

14 Repeat the process of adding a field for the following three fields:

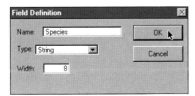 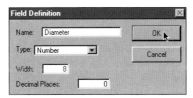

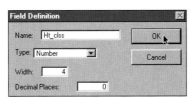

Now you will enter data about the tree you digitized in each of the table's attribute fields. It is a red maple (RM), has a diameter at breast height of 21 inches, and has a height greater than 45 feet. All trees with a height greater than 45 feet are assigned to class 3.

15 Click the Edit tool and click in the Tree_id field of the table you just created. Type **1** and press Tab. In the Species field, type **RM** and press Tab. In the Diameter field, type **21** and press Tab. In the Ht_clss field, type **3** and press Enter. The attribute table should look like this:

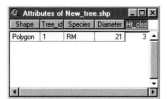

Now that you have information for tree 1 in the attribute table, you are ready to digitize more trees and perform more data entry.

16 Return to the View window and put your cursor on the base of the next tree north of tree 1. Drag a circle with a 22-foot radius (this tree's measured crown radius).

17 Return to the attribute table and enter the following data in the appropriate fields for tree 2.

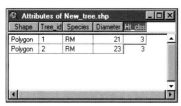

18 Repeat steps 16 and 17 to create a 22-foot circle that represents the tree north of tree 2. Enter the following attributes for this tree:

Tree_id: **3**   Species: **WO**   Diameter: **30**   Ht_clss: **3**

Press the Enter key on the keyboard.

19 From the Table menu, click Stop Editing. Click Yes to save edits to New_tree.shp. Click the Select None button.

19a Identify two ways that tree 3 differs from the other two trees.

20 Close the attribute table. Make School Property the active theme and click the Zoom to Active Theme button.

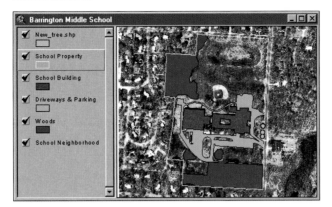

From this perspective, you can easily see the three trees you digitized. Now that you know how to digitize trees on a basemap and how to create and populate an attribute table for a tree inventory, you will explore and practice the analysis of the completed tree inventory. In the interest of time, you will not enter the data for 217 trees. Instead, you will add a tree theme with all of the data already entered.

21 If you need to exit ArcView now, you have the option to save your project. If you are in a class-room environment, ask your instructor for directions on how to rename your project and where to save it (e.g., **tree_abc.apr** where "abc" represents your initials). You will not need this saved project to complete part 2 of the exercise.

PART 2   ANALYZING THE BARRINGTON TREE INVENTORY

1 If necessary, start ArcView and navigate to the module 7 exercise folder *(C:\esri\comgeo\module7).* Open *tree.apr* or *tree_abc.apr.*

2 Turn off New_tree.shp. Click the Add Theme button and navigate to the exercise folder *(C:\esri \comgeo\module7).* Select the theme tree.shp and click OK. Turn on Tree.shp.

3 Make Tree.shp the active theme and open its attribute table.

Shape	Tree_id	Pub_prv	Species	Dia_clss	Diameter	Ht_clss	Health	Cnf_eov
Polygon	2	B	WA	2	12	3	2	N
Polygon	3	B	WA	3	31	3	3	N
Polygon	4	B	RO	3	50	3	4	N
Polygon	5	B	GM	2	11	2	3	N
Polygon	6	B	GM	2	15	2	3	N
Polygon	7	B	RO	3	56	3	3	N
Polygon	8	B	WA	2	13	3	2	N
Polygon	9	B	BOK	3	61	3	4	N
Polygon	10	B	CH	1	9	3	4	N
Polygon	11	B	WA	3	39	3	3	N
Polygon	12	B	BOK	2	17	3	3	N
Polygon	13	B	BOK	2	16	2	3	N

Notice that the most significant difference between the attribute table for Tree.shp and the attribute table you created in part 1 for New_tree.shp is that there are more fields and records in this table. The field definitions for the Tree.shp attribute table are given below.

FIELD NAME	FIELD DEFINITION
Tree_id	Tree identification number
Pub_prv	Ownership Code: A = Private; B = Public; C = Unknown
Species	2–3 letter code for tree species
Dia_clss	Diameter class: 1 = < 10 in.; 2 = 10–20 in.; 3 = > 20 in.
Diameter	DBH (Diameter at Breast Height) in inches
Ht_clss	Height class: 1 = < 25 ft.; 2 = 25–45 ft.; 3 = > 45 ft.
Health	Code for tree health: 0 = Unknown; 1 = Dead/dying; 2 = Poor; 3 = Fair; 4 = Good; 5 = Excellent
Cnf_eov	Conflicts with existing overhead utility lines: Y or N or U (unknown)
Cnf_str	Conflicts with adjacent structures: Y or N or U (unknown)
Cnf_side	Conflicts with curb or sidewalk: Y or N or U (unknown)
Cnf_tree	Conflicts with adjacent trees: Y or N or U (unknown)

3a What is the ownership and health of tree 5?

3b Of trees 2–10, how many have conflicts with overhead utility lines or adjacent trees?

4 Select the Species field by clicking on the field name at the top of the column. It will turn darker gray. Click the Sort Ascending button to list the tree species in alphabetical order. Refer to the table below for the eight most populous tree species codes and the species names.

SPECIES CODE	NAME OF SPECIES
BOK	Black oak
HEM	Hemlock
PO	Pin oak
RM	Red maple
RO	Red oak
WO	White oak
WP	White pine
YEW	English yew

5 Click the Pointer tool. Scroll down the list until you see the RO entries. Hold down the Shift key while you click each of the red oaks in the table. (Shortcut: Hold down the Shift key, then click on the first RO row and drag your mouse to the last RO row.)

5a How many red oaks are there in this inventory?

✷ **HINT:** Look at the status bar above the table to see how many records are selected.

6 Close the attribute table and return to the View window.

6a What do the yellow highlighted trees represent?

6b What is their distribution around campus?

7 Click the Clear Selected Features button to clear the selected trees.

Now you will perform a query that identifies all the mature trees in the inventory. They have a Diameter class of 3 (greater than 20 inches).

8 Click the Query Builder tool. In the Fields list, double-click [Dia_clss]. Click the equals button (=) once. In the Values list, double-click 3. Click New Set.

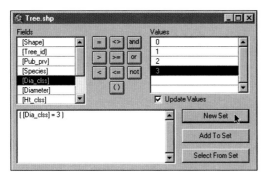

9 Close the query dialog box and look at the map.

9a What can you conclude about the age of the tree population on the Barrington Middle School property?

9b List three questions about Barrington Middle School's tree population that you could answer and analyze with this data. If you have time, try to answer your questions.

✳ **HINT:** Clear selected trees if necessary.

Your next task in this exercise is to identify trees that are potentially hazardous to utility lines or the school building itself. The most dangerous trees are ones that conflict with utility lines or the school building and also are in poor condition. A tree in poor condition (or one that is dead or dying) is much more likely to fall or lose major limbs in a storm. With your map, the Public Works Department will be able to remove the dangerous limbs before they cause damage.

Next, you will use ArcView software's query function to identify trees that conflict with overhead utility lines (Cnf_eov = Y).

10 Clear any selected trees. Click the Query Builder tool. Set up the query as follows: From the Fields list, double-click Cnf_eov; click the equals button (=) once; from the Values list, double-click Y.

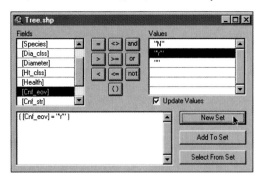

11 Click New Set. Move the Query Builder window so you can see the map and close the attribute table if it's still open.

11a What has changed on the map?

11b What do these changes mean?

Now you will identify trees that conflict with the school building (Cnf_str = Y).

12 Click the Query Builder window. Delete the first query from the text box. Set up the next query as follows: From the Fields list, double-click Cnf_str; click the equals button (=) once; from the Values list, double-click Y. Click Add to Set. Move the Query Builder window so you can see selected trees.

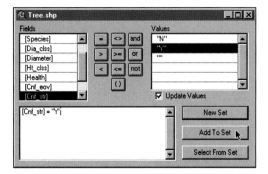

12a How did the map change?

The selected trees on your map represent all the trees having conflicts with either the utility lines or the school building. Now you will identify all of the selected trees that are in poor or dead or dying condition (Health <= 2).

13 Click the Query Builder window. Delete the previous query. The new query should be set up as follows: From the Fields list, double-click Health; click the less than or equal button (<=) once; from the Values list, double-click 2. Click Select From Set.

Close the Query Builder window.

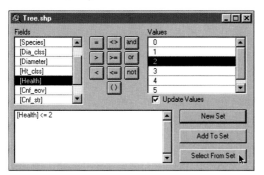

The selected trees are the hazardous trees—the ones in poor condition (or dead or dying) that conflict with utility lines or the school building.

13a How many of the middle school's 217 trees are hazardous?

✷ HINT: Look at the attribute table.

13b Is there any pattern to the location of the hazardous trees? If so, where are they?

Now that you have selected all of the hazardous trees, you will create a new theme that represents these trees.

14 Make sure that Tree.shp is the active theme. From the Theme menu, click Convert to Shapefile. Navigate to a folder where you can save the new theme. Name it **hazard_abc.shp** where "abc" represents your initials. When prompted to add the new shapefile to the view, click Yes.

15 Clear the selected features in the theme Tree.shp. Turn on Hazard_abc.shp. If necessary, use the Legend Editor to change the color of Hazard_abc.shp to one that contrasts with Tree.shp.

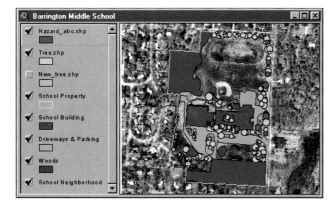

16 Change theme name Hazard_abc.shp to **Hazardous Trees** and change Tree.shp to **Other Trees**.

Now you are ready to create a printed map for your presentation to the Public Works Department that displays the hazardous trees. ArcView makes it easy to prepare presentation maps called *layouts*.

17 Be sure that all themes are turned on except New_tree.shp. From the View menu, select Layout. Click OK for Landscape. Enlarge the Layout1 window so you can see the layout more clearly.

18 Double-click the map title (Barrington Middle School) and change it to **Hazardous Trees: Barrington Middle School**. Reposition and resize the title so it's centered at the top of the layout. Reposition the legend and the north arrow so there's space for your name and date on the right side of the layout.

Community Geography: GIS in Action

19 Click the Text tool and click the layout where you want your name and date to go. Complete the Text properties window and click OK. Look over your presentation layout and make all final changes. Your layout is now ready to print.

20 Print your map.

21 Save your project. If you are in a classroom environment, ask your instructor for directions on how to rename your project and where to save it (e.g., **tree_abc.apr** where "abc" represents your initials). If you are not going to save the project, exit ArcView by choosing Exit from the File menu. Click No when asked if you want to save changes to tree.apr.

SUMMARY

In this exercise, you:
- Learned how to digitize trees from an aerial photograph
- Entered attribute data for features
- Queried the tree data to identify hazardous trees
- Created a layout of the hazardous trees

ASK A GEOGRAPHIC QUESTION

Urban sprawl and unplanned growth affect community environments—and particularly community trees—everywhere. Whether you live in an urban, suburban, exurban or rural community, the chances are that your area has been affected by this widespread problem. There is a universal need for long-term plans that will enable us to preserve and protect our community forests as our cities and towns grow and develop in the future. It's ultimately a question of sustainable development—how can we manage our arboreal resources today so that future generations will continue to experience the beauty and the benefits that community trees provide?

Before you can plan for the future, it is essential to know what the status of the community's tree population is today. The fundamental question underlying any tree inventory is "What is the size and condition of our community forest today?" In the inventory, that question will be subdivided into questions about the species' diversity, age, health, and potential problems in the study area's tree population. Town planners and maintenance workers can develop a long-term plan of managed care of the community's "green umbrella" after these questions are addressed.

Select your study site

Your first important task in preparing a tree inventory is selecting a study site. Although the ultimate goal is to record data for the community's entire tree population, it is best to begin with a discrete and clearly defined study area. Over time, the inventory can grow to include multiple study sites, but that will be accomplished one step at a time. The initial tree inventory is the most important because it defines the database that will be used for collecting future tree data in the community. It also sets the protocols and standards for future data collection. Among the possibilities to consider for your initial inventory study site are school grounds, parks, recreational areas, or municipal properties such as town hall or public library. A public area is best because it eliminates questions of legal access to the property.

ACQUIRE GEOGRAPHIC RESOURCES

The two critical data sets that you need in order to conduct a similar project in your own community are tree inventory data and a digital orthophotograph. The tree data is the information about each tree that is collected in the field during the actual tree inventory. The digital orthophotograph (DOQ) is a georegistered aerial photograph of your study site that will be used to create a basemap in your GIS application.

Possible community issues to investigate with a tree inventory

➤ Potentially hazardous trees

➤ Species diversity of the urban forest

➤ Age distribution of the urban forest

➤ Locations that need trees

➤ Prevalence or spread of a particular tree disease or pest

➤ Improvement of composition of desired trees (e.g., drought-tolerant versus water-intensive trees, native versus introduced species)

TYPE OF DATA	ITEMS TO CONSIDER
Digital orthophotography of your study site	• Depending on ownership and copyright issues, you may have access to the data without charge or you may have to pay a fee to acquire it.
	• Contact your town's municipal government to find out where you can acquire a digital orthophotograph.
	• Contact any GIS user in your town or region of the state (utility companies, phone company, and so on) for available DOQs.
	• Two excellent sources for DOQs online are *www.usgs.gov* and *www.terraserver.microsoft.com*.
	• Contact your town, county, or state's planning department.
	• Make sure to acknowledge your data source in any publication or presentation.
Tree inventory data	• Determine the content and characteristics of your inventory database by talking to urban foresters and municipal officials who will use your data.
	• Work with municipal planning and maintenance departments to clarify and define your database.
	• Prepare a data collection protocol so that all inventory participants are following the same procedures, using identical equipment, and using the same criteria for evaluating trees.
	• Set up a data-recording sheet to match the data fields you will create in the GIS project.
	• Hold a training day for all inventory participants before actually beginning the inventory.

There are many sources of information on how to conduct a tree inventory. If possible, it is a good idea to participate in a tree inventory elsewhere with experienced foresters and arborists before undertaking one of your own. Check the Community Geography Web site *(www.esri.com/communitygeography)* for national organizations that have guidelines and programs for tree inventory planning.

Explore the possibility of developing partnerships with local organizations or groups in the implementation of your inventory. Local resources for information on inventory procedures include the following:

• State or county foresters in the department of environmental management or department of natural resources

• Local tree organizations and tree societies

• Parks, arboretums, and botanical gardens

• Professional arborists

EXPLORE GEOGRAPHIC DATA

Technical issues

There are a number of technical issues to consider in employing GIS to map and analyze a tree inventory.

- Obtain the highest resolution DOQ that is available. Resolution refers to the pixel size of the image, and is an indication of how much detail is discernible. A 1-meter pixel will enable you to see individual trees, but a 10-meter pixel will not.

- Create the table in a spreadsheet program and join the completed table to your ArcView project. To do this, digitize the tree theme on your basemap and assign each tree the appropriate ID number in the attribute table. Then, add the spreadsheet table to your project. Finally, use the common field (ID) to join the table to the tree attribute table.

Preliminary data exploration

Before beginning your actual analysis of the tree inventory, take time to explore the inventory data and familiarize yourself with its content, characteristics, and potential spatial patterns.

Suggestions for preliminary exploration:

- Use the attribute table to explore the diversity of tree species on the study site. Select individual species in the table to observe their distribution in the site view.

- Use the Legend Editor to symbolize the tree data in a variety of ways: unique value for diameter class (an indicator of age), unique value for condition, and so on.

- Make copies of the tree theme in the table of contents and then use the Table Properties definition function to define new themes based on your data. For example, you can make a theme for all trees that are the same species or the same maturity.

- Use the Query Builder to select trees that reflect specific combinations of attributes: trees that are mature and in excellent condition, that are mature and in poor condition, that conflict with utility lines and the sidewalk or curb, and so on.

- Create a buffer around buildings to see how many trees are within different distances of study site structures.

- Try to identify potential problems that could be analyzed with your inventory data in terms of species diversity, tree health, tree age, and tree hazards.

ANALYZE GEOGRAPHIC INFORMATION

Tree inventory analysis varies according to the user's needs and emphasis, but typically a tree inventory report answers the following questions:

- What is the species diversity of the tree population in the study area?
- What is the age structure of the tree population in the study area?
- How healthy is the tree population in the study area?
- Are there any trees in the study area that pose a risk to people or property?
- What are the long-term and short-term tree-care needs in the study area?

Use GIS to answer these questions. The approach used by the Barrington Middle School students was to create a separate view to analyze each of the questions. This made it easier to prepare layouts that illustrate conclusions.

What is the species diversity of the tree population in the study area?	• Symbolize the tree theme with a Unique Value legend classified by species. This enables you to see the range of species on the site and a count of each species. • Sort the attribute table by species to identify the range of species represented by the data. • Use Theme Definition to select all trees of the dominant species in the attribute table. • Create new themes for all dominant species in the inventory.
What is the age structure of the tree population in the study area?	• Symbolize the tree theme with a Unique Value legend classified by diameter class (a commonly used indicator of age). • Use Theme Definition to create a theme for a particular diameter class.
How healthy is the tree population in the study area?	• Symbolize the tree theme with a Unique Value legend classified by health class. Assign different colors for each health class. • Use Theme Definition to create a theme for a particular health class.
Are there any trees in the study area that pose a risk to people or property?	• Before answering this question you must decide which tree attributes or combination of attributes might represent a risk to people or property. One formula is the number of trees that are in conflict with buildings or streets or in poor health. There may be more than one formula that results in a "hazardous" designation. • Use the Query Builder to select all trees that are hazardous (e.g., in conflict with buildings or streets or in poor health).

Drawing conclusions
The key to drawing conclusions is in being a critical and knowledgeable observer of the data as it is mapped and queried. The following lists provide examples of different maps that could answer each question.

What is the species diversity of the tree population in the study area?
- Indigenous trees versus species that have been introduced
- Exotic trees and ornamentals
- Evergreen versus deciduous trees
- Presence and identity of dominant species
- Trees known to be endangered by pests or diseases
- Patterns of species distribution within the study area
- Advantages of species diversity
- Recommendations for species to be planted in the future

What is the age structure of the tree population in the study area?
- Rates of growth in different species
- Environmental conditions affecting growth rates
- Overall youth or maturity of the trees in the study area
- Size of mature population (percent of total) that will likely need to be replaced in the future
- Need for managed care of both young and old trees to ensure longevity

How healthy is the tree population in the study area?
- Environmental factors that affect health
- Extent of impervious surfaces in the study area
- Percent of study-area trees in each health category
- Presence of disease or pests
- Variations among species

Are there any trees in the study area that pose a risk to people or property?
- Proximity of trees to buildings, parking areas, utility lines
- Evidence of sidewalk or road "lifting" from tree roots
- Possibility for structural failure due to poor health
- Potential for remediation

As you develop conclusions and recommendations based on your analysis, prepare layouts, graphs, and data tables to illustrate your ideas. These components can be included in an inventory report document and in multimedia presentations about the benefits of trees, the importance of community forestry planning, and the process of conducting a tree inventory.

ArcView CITYgreen extension

CITYgreen is a community forestry application for ArcView developed by American Forests. It automates the process of using GIS to map and analyze a tree inventory. It has many built-in modeling, statistical, and analytical functions, is easy to use, and significantly expands the ability of community groups to analyze tree inventory data. Using CITYgreen, they can calculate the economic and environmental benefits of a tree population, model changes in a tree population and its benefits over time, and compare the benefits of alternate scenarios for the same site. It is a valuable tool for community planning. Refer to the Community Geography Web site *(www.esri.com/communitygeography)* for more information.

ACT ON GEOGRAPHIC KNOWLEDGE

After analysis, it is time to put this valuable information to use. The information you have collected is vital to public officials for the development of a comprehensive tree management program, a planting plan for future replacement of older trees, and appropriate tree ordinances and legislation. It is an essential instructional tool to teach people about one of the most important resources in their community. It is a baseline for evaluating the success of community forestry initiatives in the future.

Possible action steps

- Prepare a tree inventory report document, including recommendations suggested by your conclusions, and present it to town officials and appropriate property owners (school officials, cemetery owners, park managers, and so on).

- Prepare a multimedia presentation to give in a public forum to raise awareness of and generate support for long-term community forest planning.

- Educate others about the importance of long-term community forest planning by volunteering to give presentations in schools, to service groups (e.g., Kiwanis, Rotary), and to youth groups (e.g., scouts, church groups).

- Organize a local tree society to continue the tree inventory work you have begun and to advocate for comprehensive tree management planning.

- Expand the tree inventory into other areas of the community. Involve other local organizations in the data collection.

- Look into federal and private grant programs that provide money to support community forestry initiatives at the local level.

Next steps

The completion of a tree inventory is not the end of a project, but the beginning of a journey on the road to sustainable development for your community. In communities where citizens have initiated tree inventories, one sees a growing concern for the community's "green infrastructure" by members of conservation committees, planning boards, and town councils. An active tree lobby educates the public about the vital importance of becoming stewards of our "green umbrella" as we plan for future growth and prosperity.

MODULE 7 ACKNOWLEDGMENTS

Thanks to the students of Barrington Middle School and Lyn Malone for contributing this module's case study.

Aerial photograph of Barrington provided courtesy of New England Gas Company and is used with permission.

Barrington Middle School tree data provided courtesy of and copyright © 2000 Barrington Middle School and is used with permission.

Barrington Middle School feature data provided courtesy of Lyn Malone and is used with permission.

Selecting the right location

"Location, location, location." It's the most important factor in determining the best house to buy, the best building to rent for a successful business, or the best land parcel to preserve for wildlife habitat. Individuals, companies, and governments invest significant time and resources in determining what makes the best location. Usually, a number of criteria must be considered—some of them requirements, and others, preferences. A city might require that a house not be built in an area that could flood; a business might prefer to be in a location convenient for its customers.

Determining the specific locations that best meet all the criteria can be a daunting task. More and more, people are using GIS to sift through many layers of data reflecting the criteria. Maps can be created that show all the sites meeting the criteria. If criteria change or are refined, the GIS site selection analysis can be quickly repeated to select new candidate sites.

CASE STUDY

Using site analysis to develop a wildlife area management plan

Read about a group of eighty ninth graders at Steamboat Springs High School who worked with the Colorado Division of Wildlife to draft a management plan for a new state wildlife area. Students used GIS and GPS technologies to tackle this comprehensive project and to site parking lot options for the wildlife area.

EXERCISE

Perform site selection for a state wildlife area

Use the same data collected by the Steamboat Springs students and the same criteria for parking lot locations to propose a third parking lot option. You will add data, create buffers, and conduct spatial analysis to make your parking lot recommendation.

ON YOUR OWN

Learn about the various issues surrounding site analysis and get tips on how to perform a site analysis in your community. Information on how to form community partnerships and acquire appropriate data will be addressed.

Using site analysis to develop a wildlife area management plan

Routt County, Colorado

When Libbie Miller of the Colorado Division of Wildlife (CDOW) suggested that a group of ninth graders be responsible for drafting a management plan for a newly acquired state wildlife area, most people thought it would never get done. With the help of The Orton Family Foundation and many community partners, eighty students from the Steamboat Springs High School FLITE (Freshmen Learning In a Team Environment) program spent the 1999–2000 school year making history with their comprehensive management plan.

Libbie Miller of the Colorado Division of Wildlife (CDOW) gives ninth-grade students an on-site orientation to the Chuck Lewis State Wildlife Area. Students spent the 1999–2000 school year collecting numerous data sets on this land and drafted a wildlife area management plan.

PROJECT OVERVIEW

The overall project was ambitious and extensive. Teachers and community planners began work during the summer of 1999 and the students collected data and performed analyses throughout the following school year. One of the many goals of the project was to find a parking lot site for the new Chuck Lewis State Wildlife Area. Using a team approach, along with GIS and GPS (Global Positioning System) technologies, students completed four project components: data collection and analysis, written research, oral presentations, and visual presentations like maps and posters.

The Orton Family Foundation and the Colorado Division of Wildlife (CDOW) partnered with the FLITE (Freshman Learning In a Team Environment) class at Steamboat Springs High School to draft a wildlife management plan, including site analysis for a parking lot. The Orton Family Foundation is a nonprofit organization that supports youth, educators, and community partners working to solve local problems in their communities using GIS and GPS technologies.

THE ORTON FAMILY FOUNDATION

SITING A PARKING LOT: WHERE TO BEGIN?

Two of the student teams were assigned to determine where to locate a parking lot for recreational access within the wildlife area. The students called upon the following community experts to guide them in their background research and help them formulate relevant questions:

- County planner: For capital improvement policies and regulations
- GIS specialist: For help with ArcView applications and analyses
- Naturalists: For help identifying critical riparian habitat
- CDOW project mentor: For overall guidance and agency policy concerning roads, rivers, trails, wetlands, parking lots, and wildlife habitat

After extensive research and meetings with their community mentors, the student groups generated a list of secondary questions and answers:

WHAT SIZE LOT WOULD ACCOMMODATE 10–12 VEHICLES?	The lot should be approximately 100 by 150 feet (15,000 square feet) in size.
WHAT AREAS SHOULD BE AVOIDED?	Wetlands, setbacks, historical structures, critical wildlife habitat, and unstable or hazardous areas should be avoided.
HOW VISIBLE FROM THE ROAD SHOULD THE PARKING LOT BE?	Allow for early sighting of lot while driving; allow for safe exit and entry into traffic; avoid ridgelines and hidden valleys.
WHICH OF THE TWO COUNTY ROADS, CR14 OR CR14F, WOULD PROVIDE SAFEST ACCESS?	There is no preference, but the entrance should not be near the intersection of the two roads. CR14 has a long, straight section on a hilltop worth consideration.
WHERE ARE THE CLOSEST TRAILS THAT PROVIDE WALKING ACCESS TO THE RIVER OR OTHER TRAILS?	A few trails start at CR14F near the bridge; a connector trail may need to be built from CR14.
WHAT COUNTY AND AGENCY POLICIES MIGHT AFFECT THE LOCATION OF THE LOT?	Wildlife area boundary setback = 100 feet. River setback = 300 feet minimum; 500 feet is ideal to minimize environmental impact.

GATHERING THE NECESSARY DATA

With their guiding questions answered, students were ready to gather the necessary data. For the basemap, CDOW provided a map of the wildlife area and adjacent landowner boundaries, and the local GIS specialist gathered digital and aerial photographs from Steamboat Springs and Routt County. With guidance from the teachers, CDOW, and community mentors, the students documented many key landmarks within the wildlife area with GPS units and digital cameras. The data collected by students included internal and adjacent county roads, fence lines, bridge location, ditches, a headgate, an old pump house, old farm equipment, a shed, hay fields, an underground electrical supply, and an old corral.

Throughout the year, students used GPS technology to record the latitude and longitude points of structures and landmarks in the wildlife area, such as abandoned farm equipment (left) and a run-down pump house (right) in the wildlife area.

CREATING INFORMATIVE MAPS AND SURVEYING PUBLIC OPINION

The student groups were required by their teachers and the CDOW to conduct three open houses to gather community feedback on the use of the wildlife area. To prepare for the open houses, creating informative maps was at the top of the students' to-do list. First, they wanted to display a map showing all of the existing land features.

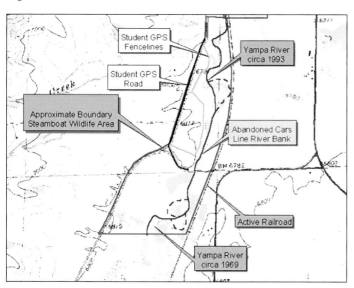

Advanced GPS data collection and GIS skills were necessary to create this map showing the entire wildlife area, the Yampa River location in 1969 and 1993, roads, fences, and the railroad.

When students presented their maps and surveys to the public at the open houses, the public expressed interest in using the wildlife area for recreational activities such as hiking, horseback riding, and waterfowl hunting. This feedback confirmed the perceived need for a parking lot that would hold 10–12 vehicles, including a few horse trailers.

After collecting this valuable information, the students used GIS software to create a spatial framework from which they could determine a parking lot location. Although the students had a topographic map showing the Yampa River, the river's course had changed significantly since 1969 when the topographic map was published. The GIS specialist helped the students trace the river's more recent location from a 1993 aerial photograph. This data was added to a map with data on roads, topography, wildlife boundary, and structures.

With all their data in the map, they created the buffers based on CDOW's zoning requirements and added them as data layers to their map:

- 300-foot buffer around the river inside the wildlife area (CDOW requirement)

- 500-foot buffer around the river inside the wildlife area (CDOW preferred setback)

- 100-foot buffer inside the entire wildlife area boundary (CDOW requirement)

CDOW told the students that the parking lot had to avoid the land within each buffer, avoid other structures (such as the old pump house and corral), and be located next to a safe section of county road. The students found that the land within the buffers significantly limited the available land for parking lot locations. In addition, they studied hard-copy floodplain and geologic hazards maps provided by the county. Their analysis led them to focus on two possible locations.

After creating the 500-foot buffer around the river (displayed with diagonal pink lines) and the 100-foot buffer inside the wildlife boundary (displayed as a thick pink area), students focused on the available area east of the road (labeled as Student GPS Road).

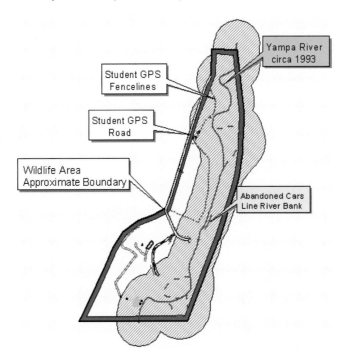

214

The students drew a 100-by-150-foot rectangle in ArcView to represent the dimensions needed for the parking lot. They duplicated it so they could see how different lot locations graphically fit in relation to other mapped features. After visually analyzing the map, they settled on two possible parking lot locations.

PROJECT RECOMMENDATIONS

During the three public open houses, the students presented their proposed parking lot locations and obtained feedback from citizens. The public supported the students' first choice, option 2 off CR14, just north of the intersection with Elk Lane.

At three open houses held throughout the 1999–2000 school year, students gave presentations and solicited community member feedback through surveys. The FLITE group obtained more public input than the state wildlife agency had historically been able to accomplish through traditional outreach methods.

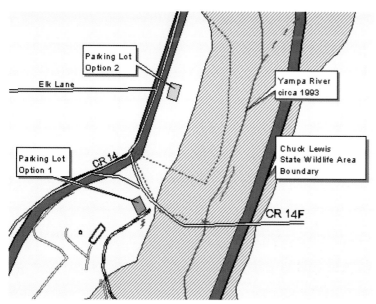

The students determined two possible locations for a parking lot and mapped them. They presented these options at the public open houses and obtained feedback. At the end of the project, they recommended option 2 to the Wildlife Commission.

The FLITE team presented the benefits and disadvantages of both parking lot options, and recommend parking lot option 2.

PARKING LOT OPTION	BENEFITS	DISADVANTAGES
1	• Close to the river for easy access • Off of busy county road	• In the floodplain • Negative impacts to wetlands area • In an agricultural field
2	• Can serve as access for the nearby CDOW property on Emerald Mountain • Out of the floodplain • Outside any areas with geologic hazards	• Sight distance limited to a blind corner at the junction of CR14F and CR14

The students concluded their project in June 2000 by presenting their final Chuck Lewis State Wildlife Area Draft Management Plan, including their comprehensive recommendations for best management approaches, to the Routt County Commissioners and the State Wildlife Commission. They gave the state wildlife management agency the ArcView basemap of the wildlife area to use in future planning. The students and community members alike also gained an appreciation of the wildlife agency's public service role, its dependence on public input for formulating the best management policies, and the importance of citizens becoming involved in setting policies that will ultimately affect them and their community.

A student answers questions from the Wildlife Commission about the Chuck Lewis State Wildlife Area Draft Management Plan (left). Wildlife commission members listen to students present their recommendations (right).

At the conclusion of their presentation, Commission Chairman Bernard Black recognized the value in having students involved in public lands management issues with these words:

"I am very impressed with the work you have done. I think it is so important for young people to get a handle very early on conservation and the environment and the things one needs to do to keep our habitats in good enough shape to support our wildlife. . . . This is a unique program that is beneficial to the environment and the community. I'd like to see it expand to other parts of the state."

WHAT HAS BEEN DONE?

The management plan has since been used by CDOW wildlife managers to develop and manage the property for wildlife and wildlife-related recreation, including parking lot construction. Although their school project has been complete for over two years, many students are still seen participating in other public meetings on conservation issues.

Student recommendations were realized—the parking lot, information kiosk, and signs were all built according to the standards outlined in the Chuck Lewis State Wildlife Area Draft Management Plan in summer 2000.

The parking lot analysis was only a small part of a much larger project. For many months the students on the FLITE team worked in partnership with the Colorado Division of Wildlife and The Orton Family Foundation to draft a management plan for the Chuck Lewis State Wildlife Area. The following table summarizes the geographic questions the students answered when researching and writing the draft.

GEOGRAPHIC QUESTION	STUDENT TEAMS	ISSUES
Should the line of abandoned cars be removed from along the bank of the Yampa River? 	• Structures • Fisheries Management • Aquatic and Riparian Management	• The location of the abandoned cars using GPS and from what points in the wildlife area the cars can be seen • How the cars affected the riverbank and native trout and nonnative pike populations • Environmentally safe bank stabilization options that would replace the cars • How to work within CDOW and county policies and public values to evaluate the aesthetic, historic, and environmental impact of the abandoned cars
How big of a parking lot is needed and where should it be located? 	• Structures • Aquatic and Riparian Management	• Lot size needed to accommodate ten to twelve cars • Areas that should be avoided due to CDOW rules on minimum setback, historical structures, wildlife habitat, and hazardous areas • Which county road provides the safest access
Which areas of the wildlife area should be closed seasonally or permanently?	• Closed Areas • Seasonal Use Areas	• Areas used for bird nesting that should be recommended for seasonal closure • Unsafe areas that should be closed for public safety • Which areas should be closed to protect wildlife habitats • Which areas should be closed to small- and big-game hunting and when

GEOGRAPHIC QUESTION	STUDENT TEAMS	ISSUES
How many and what kinds of regulatory and interpretive signs should be installed? 	• Signs/Portal • Mapping Structures	• Which signs and how many are needed to regulate public use • What interpretive messages and signs would help the public appreciate the wildlife area • Where signs should be placed, such as at portals or entry points
What improvements should be made to the riparian areas? 	• Aquatic Habitat and Riparian Management • Fisheries Data and Management • Threatened and Endangered Species Data	• Areas of the stream bank that should be revegetated or stabilized to control erosion • Which wetlands areas should be enhanced for wildlife habitat or to improve water availability • Where and how many fish habitat structures should be installed along the river
What improvements should be made to the terrestrial areas? 	• Terrestrial Species Data • Rangeland Management • Internal Fences • Structures • Roads/Trails • Ditches/Head Gates • Power Lines/Pipelines	• The location of existing structures and whether these should be removed • Where agricultural practices could be sustained on the property • Whether CDOW service roads should be improved and/or installed • Where new fencing is needed on the property either as enclosures or exclosures • Should any structures be preserved for their historical value?

SUMMARY

ASK A GEOGRAPHIC QUESTION	• What size parking lot is needed? • What areas should be avoided? • Where should the lot be in relation to the road, the wildlife area boundary, and the Yampa River? • Where are the closest trails to provide access from the parking lot to the river? • How do county or agency policies influence the lot location?
ACQUIRE GEOGRAPHIC RESOURCES	• Obtain map of wildlife area and landowner boundaries from CDOW. • Work with local GIS specialist to obtain digital quadrangles and aerial photographs and other feature maps from the city and county. • Collect the following data in the field: location of human-made structures on the property, ditches, roads, hay fields, and natural features. • Record data in a table that can be imported into ArcView.
EXPLORE GEOGRAPHIC DATA	• Create a buffer around the river and within the wildlife boundary to determine where the parking lot cannot be located. • Visually analyze the available space and how it conflicts with other human-made structures that cannot be removed.
ANALYZE GEOGRAPHIC INFORMATION	• Create a rectangular area representing 100 by 150 feet to place in different locations on the map to reflect the parking lot size needed. • Identify probable locations based on buffers, preexisting structures, proximity to the road, and safety. • Create maps that display the two options for parking lots.
ACT ON GEOGRAPHIC KNOWLEDGE	• Summarize your recommendations. • Prepare presentations and practice them. • Solicit public opinion at three open houses. • Incorporate these recommendations into the draft management plan. • Give presentation to Wildlife Commission. • Deliver database of information and maps to the local division of wildlife.

Perform site selection for a state wildlife area

In this exercise, you will propose a third option for a parking lot location. You will work with much of the data that the Steamboat Springs High School students used. You will use their 100-foot buffer from the wildlife area boundary, but you will need to create your own 300- and 500-foot buffers for the Yampa River. You will also draw a 100-by-150-foot polygon that you will move around, evaluating different locations and avoiding features such as waterways and setbacks. Once you have settled on the location for option 3, you will save the polygon as an addition to the parking lot shapefile.

The ⊟ icon indicates questions to be answered. Write your answers on a separate sheet of paper.

ASK

The Steamboat Springs High School students came up with two potential sites for the 10- to 12-car parking lot that serves the new Chuck Lewis State Wildlife Area. The Colorado Division of Wildlife (CDOW) would like to present three options at the next public open house. They have asked you to evaluate the data and come up with a third potential parking lot site.

CDOW has specific criteria about where the parking lot can be located:

- It needs to be at least 300 feet away from the river and inside the wildlife boundary. Ideally, it will be 500 feet or more from the river.
- It needs to be at least 100 feet inside the wildlife area boundary.
- It has to avoid other structures (such as the old pump house and corral) and be located next to a safe section of road (CR14, CR14F, or Elk Lane).

Given these guidelines, you need to determine the best site for a third parking lot option. Consider the advantages and disadvantages of this site and how it compares to the other sites the students chose.

EXPLORE

1 Start ArcView. From the Open Project window, navigate to the exercise data folder *(C:\esri \comgeo\module8)*. Open the project *wildlife.apr*. If you receive an Update project dialog, click Yes to add the new tools.

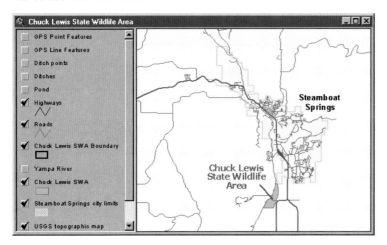

When the project opens you will see a map showing the Chuck Lewis State Wildlife Area (Chuck Lewis SWA) located to the south of the city of Steamboat Springs.

2 Zoom in to the wildlife area so that the map scale (displayed above the view) is larger than 1:25,000.

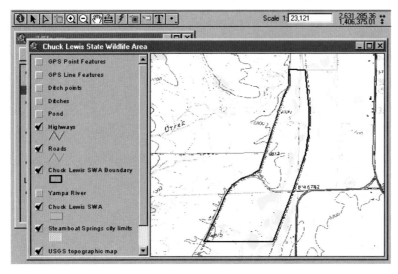

Notice that the themes displayed in the view changed when you zoomed in. Several of the themes in this view have scale dependence so they display only at an appropriate map scale. For example, the green polygon from the Chuck Lewis SWA theme disappears when the map scale is larger than 1:25,000, and the black outline of the Chuck Lewis SWA Boundary theme appears instead.

The view contains a number of other themes of data created by the Steamboat Springs students. You will take a few moments to explore these themes.

3 Press the Shift key and click the Yampa River theme and the five themes above Highways in the table of contents to make them active.

4 With all six themes active, choose Hide/Show Legend from the Theme menu. The theme legends are displayed.

5 Turn these themes on one at a time and observe where the various features are on the map. Zoom and pan the map as needed.

5a What features exist along the fence?

5b Based on the information in the map, what is the purpose of the pipe in the southwest section of the wildlife area?

5c How could some of these features limit the area where the parking lot could go?

The topographic map provides important orientation information about the Chuck Lewis SWA. However, the date of the map is 1969, so the information may have changed. You will add a more recent aerial photograph to the view.

6 Click the Add Theme button. Navigate to the exercise data folder *(C:\esri\comgeo\module8)*. In the Data Source Types drop-down menu, choose Image Data Source. Add aerialphoto.tif. Turn the new theme on and then move it from the top of the table of contents to third from the bottom, just above the USGS topographic map theme.

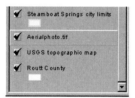

Because the aerial photograph is very detailed, it is best displayed at large map scales. You will make the aerial photograph theme scale dependent so that it displays only when the map scale is larger than 1:15,000.

7 Make sure the Aerialphoto.tif theme is active, then click the Theme Properties button. In the Theme Properties window, click the Display icon. Type 15000 in the Maximum Scale box and click OK.

8 To test the new setting, type 16000 in the Scale box on the tool bar and press the Enter key.

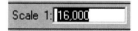

8a Why doesn't the aerial photograph display even though the theme's check box is checked?

9 Zoom in again and explore the aerial photograph. Answer the following questions by zooming, panning, and turning themes on and off as needed.

9a The Yampa River theme was digitized from the aerial photograph. Describe the changes you see, if any, in the location of the Yampa River since 1969 (as shown on the USGS map).

9b Compare and contrast the geographic information about the wildlife area that is available from the USGS topographic map, the aerial photograph, and the point and line themes. Use specific features as examples of your observations.

9c Give an example of when you would use a topographic map and when you would use an aerial photograph.

The parking lot criteria described in the exercise scenario mentions three roads: CR14, CR14F, and Elk Lane. In the next steps, you will identify these roads and label them on the map. First, you will give the Roads theme a thicker line symbol so roads can be seen on top of the aerial photograph.

10 If necessary, turn off the GPS Line Features theme and turn on the Roads theme. Open the Legend Editor for the Roads theme and change the line symbol size to **3**. Apply the change and close the Legend Editor and Pen Palette windows.

11 Zoom the map so that it is centered on the various roads running through or near the wildlife area. (Refer to the picture below.)

 12 Click the Identify tool and use it to determine which roads are CR14, CR14F, and Elk Lane. You will need to scroll down in the Identify Results window to the Name and St_id fields.

Next you will specify the font for the road labels.

 13 From the Window menu, choose Show Symbol Window, and then click the Font Palette button. Choose Arial Black, size 18. Click the Color Palette button. From the Color drop-down list, choose Text, and click the brown color box.

 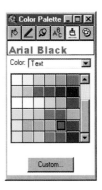

14 Close the Color Palette window. Click the Label tool. Click each road once to label it.

By default, the Label tool uses the information in the Name field. For Elk Lane, this field contains the code SP1. You will need to change the label.

15 Click the Pointer tool. Double-click the label, SP1. In the Text Properties window, delete SP1 and type **Elk Lane**. Click OK.

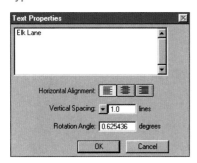

16 Turn off the Aerialphoto.tif theme and, if necessary, turn on the USGS topographic map theme. Use the Pointer tool to select and move each road label slightly apart from the road line.

ACQUIRE

Before you can choose a site for the parking lot, you need to map the parking lot criteria CDOW determined. In this exercise, the 100-foot setback from the wildlife area boundary is represented as a buffer you will add. At this time, you will also add the location data for the first two parking lot options. At the end of the exercise, you will determine where the third option will go.

17 Click the Add Theme button. Navigate to the exercise data folder *(C:\esri\comgeo\module8)* and add the following themes: dowbnd100.shp and parking_lots.shp. Scroll to the top of the table of contents and turn on the two themes.

Before continuing, you will take a few moments to symbolize and set properties for the new themes. If you're using ArcView 3.0a, you will need to load the legends for the new themes. If you're using ArcView 3.1 or higher, they were loaded automatically, so skip to step 19.

18 For each of the two themes, open the Legend Editor and click the Load button. Load the appropriate legend file from the exercise data folder using the default settings. Close the Legend Editor when you are finished.

THEME	LEGEND FILE
dowbnd100.shp	dowbnd100.avl
parking_lots.shp	parking_lots.avl

19 For each of the two themes, open the theme properties. Using the following table as a guide, rename the themes and make them scale dependent. (Refer to step 7 for help.)

THEME	NEW THEME NAME	MAXIMUM SCALE
dowbnd100.shp	100-ft Bnd Setback	1:25,000
parking_lots.shp	Parking Lot Options	1:25,000

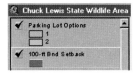

20 Drag the 100-ft Bnd Setback theme down in the table of contents to just below the Chuck Lewis SWA Boundary theme. With this theme active, click the Zoom to Active Theme button.

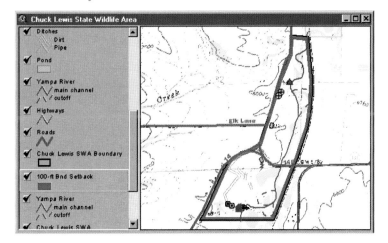

21 Save the ArcView project. If you are working in a classroom environment, ask your instructor where you should save the project. Name the project **wildlife_abc.apr** where "abc" represents your initials.

PART 2   CREATE BUFFERS AND SITE PARKING LOT OPTION 3

1 If necessary, start ArcView and open your project (wildlife_abc.apr).

In the next steps, you will use a script to create buffers representing the 300- and 500-foot setbacks from the Yampa River. At first the buffers are created as graphics, but you will copy them into shapefiles.

In order to clearly see the buffer outlines when they are added to the map, you will turn off the themes you don't need.

2 In the view, turn off all the themes except 100-ft Bnd Setback and Yampa River.

If you are using ArcView 3.1 or higher, you may want to use the Create Buffers wizard instead of the script. To do so, replace steps 3–16 with the following steps:

1 From the Theme menu, choose Create Buffers.

2 In the Create Buffers wizard, choose to buffer "The features of a theme" and select Yampa River from the drop-down list. Click Next.

3 Choose "At a specified distance" and type **500** in the box. Click Next.

4 Choose "Yes to Dissolve barriers between buffers." Choose to save the buffers in a new theme, in the folder where you are saving your work for this exercise, and name the file **buf500_abc.shp** where "abc" represents your initials. Click Finish.

5 Rename the theme **Yampa 500-ft Buffer**.

6 Repeat the steps to create a 300-foot buffer in a shapefile **buf300_abc.shp** with the theme name **Yampa 300-ft Buffer**.

7 Drag the two themes below the Yampa River theme in the table of contents.

8 Go to step 17, page 230.

Now you will create two new shapefiles, one for each river buffer.

3 From the View menu, choose New Theme. In the New Theme window, choose Polygon and click OK. Navigate to the folder where you are saving your work for this exercise, and name the file **buf500_abc.shp** where "abc" represents your initials.

4 From the Theme menu, choose Stop Editing, and click Yes to save your edits. Rename the theme **Yampa 500-ft Buffer**.

5 Repeat steps 3 and 4 to create a **Yampa 300-ft Buffer** theme.

6 Drag the two themes below the Yampa River theme in the table of contents and turn them on. (Remember, these themes contain no features yet, but turning them on now will let you see the features as soon as they are added. Your themes' default colors may be different than those shown.)

Next, you will select the features you want to buffer.

7 Make the Yampa River theme active. Click the Select Features tool and drag a box to select all of the solid and dashed blue river lines.

You are ready to create the buffers. You'll create the larger buffer first.

8 Click the Create Buffers button. (This special button runs the buffer script.)

9 In the Buffer window, enter a buffer distance of **500** feet and click OK. Click Yes to dissolve adjoining buffers.

The outline of the 500-foot buffer is added to the view as a graphic. You will copy this graphic into the 500-foot buffer shapefile.

10 Make the View window active. Click the Pointer tool, then click the black buffer outline to select the graphic. (Square handles appear. If you do not see them, zoom out.) From the Edit menu, choose Copy Graphics.

11 Scroll down in the table of contents and make the Yampa 500-ft Buffer theme active. From the Theme menu, choose Start Editing.

12   From the Edit menu, choose Paste. The buffer becomes a polygon in the Yampa 500-ft Buffer theme.

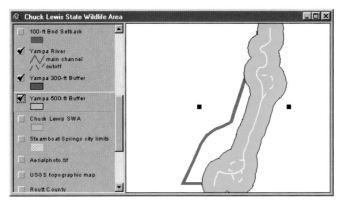

13   From the Theme menu, choose Stop Editing, and save your edits. If the buffer is selected (yellow), click the Clear Selected Features button.

You will repeat the procedure to create the 300-foot buffer.

14   Repeat steps 8–13 using a buffer distance of 300 feet and pasting the buffer into the Yampa 300-ft Buffer theme.

15   Make the Yampa River theme active and click Clear Selected Features.

Now that you have saved the buffers as shapes, you can delete the graphics.

16   Turn off the two buffer themes. Select each buffer graphic and press the Delete key.

17   In preparation for your parking lot site analysis, modify the symbols for the 300- and 500-foot buffers using fills and colors of your choice. The example below uses light- and dark-purple fill with no outlines. You might prefer to use thick outlines with no fill, or semitransparent fill patterns.

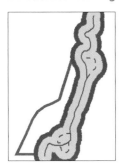

17a   Why do you think CDOW has a required buffer of 300 feet around the river and why would they prefer a 500-foot buffer?

ANALYZE

Now that you are familiar with the landscape of the Chuck Lewis State Wildlife Area and you have map layers representing the criteria for acceptable parking lot sites, you are ready to find a third parking lot site for CDOW to present to the public.

18 **From the View menu, choose Themes On to turn on all of the themes at once.**

18a What general areas do you see, if any, that would be an appropriate site for parking lot option 3?

You will add 100-year flood data for comparison with the buffers.

19 **Add the theme floodzone.shp from the exercise data folder *(C:\esri\comgeo\module8)*. Move the new theme to an appropriate place in the table of contents, and change the symbol or color for the theme if desired.**

19a How could flood-zone data help you determine where parking lots should be located?

19b How have the possible sites changed with the addition of the flood-zone theme?

20 **Zoom in and more closely examine the parking lot option 1 and 2 sites.**

20a What differences do you notice between the locations of parking lot options 1 and 2?

Next you will experiment with possible option 3 sites.

21 **Click the Draw Rectangle tool. From the Window menu, choose Show Symbol Window. Select solid fill and a color of your choice for the third parking lot option. Be sure to choose a color that will show up against the aerial photograph as well as the topographic map, such as bright red or gold. Close the Symbol Palette.**

22 **Near one of the existing parking lot options, draw a rectangle about the same size (it will have an area of approximately 15,000 square feet).**

23 **Click the Pointer tool and drag the rectangle around the map until you have settled on the best site for option 3.**

23a List the reasons why you chose this particular location for your site.

Now that you've selected the location, you will add the rectangle to the Parking Lot Options theme.

24 Make sure the option 3 graphic is selected, then from the Edit menu, choose Copy Graphics.

25 Make the Parking Lot Options theme active, then from the Theme menu, choose Start Editing. From the Edit menu, choose Paste. The rectangle is pasted as a feature in the theme.

Now you need to update the attribute table.

26 Click the Open Theme Table button, then click the Edit button. Type **3** in the Option field for the new polygon, and press Enter.

27 From the Table Menu, choose Stop Editing, and save your edits. Close the table.

The graphic is covering up the new feature in the theme, so you will delete the graphic.

28 Turn off Parking Lot Options. Make sure the Pointer tool is active, and then select the option 3 graphic and delete it.

Finally, you will update the Parking Lot Options theme legend.

29 Turn on Parking Lot Options and open its Legend Editor. Reselect Option in the Values Field drop-down list to update the legend. Choose Minerals for the color scheme. Click Apply and close the Legend Editor. Clear the selected features.

ACT

30 Compare the three parking lot locations.

30a Choose the location that best meets the criteria.

30b Explain the advantages of your chosen site over the other two site locations.

31 Create and print a layout that illustrates the locations of all three proposed parking lot sites and labels your recommended site.

SUMMARY

In this exercise, you:

• Set scale dependency so themes displayed at appropriate map scales
• Created 300- and 500-foot buffers from the Yampa River
• Visually evaluated the locations of buffers, setbacks, flood zones, and other criteria and selected a site for a third parking lot option

ASK A GEOGRAPHIC QUESTION

When developing your geographic question(s), focus on a community need and how you can meet it. Examples of places that might need a parking lot sited for them include recreation areas, a new hotel, or a school. Another possibility is to help evaluate alternative sites that have already been proposed.

Create a project plan

Site selection for a new facility or parking lot involves many different data sources and community concerns. In order to complete an effective selection process, you should be prepared to assimilate information and data from various groups in the community. Being organized will help you tremendously:

- Consider your available time and make sure the scope of your project does not exceed it.
- Plan a project time line, identifying dates by which major activities will be complete. For example, if you need to conduct a public survey as part of your project, set the dates for survey distribution and analysis.
- Develop a master plan for your project. Include the project time line as well as lists of what needs to be done by when and by whom. This master plan should include immediate, short-term, and long-term goals.
- Determine how large the site needs to be to meet community goals. Consult with the appropriate government officials and local experts for guidelines on setting a site size.
- Form community partnerships (see details below).

Form community partnerships

It is essential that you reach out to other community organizations before you begin your project. You can seek outside assistance from GIS professionals who can provide technical expertise or computer hardware and printers. Working with community organizations that have a vested interest in the success of your project is vital. They can also provide you with valuable information on local policies and laws regarding planning and zoning. In this module's case study, the FLITE students worked with the Colorado Division of Wildlife (CDOW) and The Orton Family Foundation to create a management plan for a new wildlife area. The students benefited significantly from the resources CDOW provided: data, regulatory guidelines, and a thorough understanding of the approval process. CDOW benefited from the data collection, analysis, and public outreach done by the students.

Site analysis possibilities:

- New athletic complex
- New school
- New park or playground
- Tourist information kiosks
- Video store
- Cell phone tower
- Horse trails
- Skateboard park
- Historical walking tour
- Wildlife viewing areas
- Art sculptures
- Walking trails

ACQUIRE GEOGRAPHIC RESOURCES

When you acquire geographic resources for a site-selection project, you need to focus on three areas:

1 Information about local, state, and federal laws regarding planning and zoning

2 Basemap data

3 Data collection

Before you can begin analysis, it is essential that you understand the planning and zoning laws in your community. In the case study, the FLITE students learned that CDOW had rules about how far away the parking lot could be from water sources and the boundary of the wildlife area. In addition, individual organizations may have specific criteria requirements for their site selection. For example, a warehouse distribution center may have a rule about how close it needs to be to a major highway. Check with the following organizations to learn about the required planning and zoning regulations in your area:

• State department of wildlife

• City, town, or county government (office of planning and zoning)

• Environmental Protection Agency

• Regional planning offices

• State department of transportation for traffic rules

• Community partner(s)

In addition to the regulations set by law, your group or your community partner may have preferred criteria for site selection. The FLITE students learned that although CDOW required that the parking lot be at least 300 feet from the river, the department preferred a minimum distance of 500 feet. They also learned that CDOW preferred that the lot accommodate horse trailers, even though this was not a regulation. Many businesses will have a set of preferred criteria when determining where to locate a new store. Quite often, their preferred criteria relate to the demographics of the area surrounding a new store location. Some stores will prefer a large general population size. Others will prefer a large population of a particular ethnic group or of a particular age (such as teenagers). In your project, it's important to identify the required criteria and any preferred criteria for your site selection.

Basemap data for site-selection analysis frequently needs to be very current. The following table lists the types of data you may need and suggestions as to where to find it:

TYPE OF DATA	WHERE YOU CAN LOCATE THE DATA
Demographic characteristics of your town	• Census data from American FactFinder Web site *(factfinder.census.gov)* or from *www.geographynetwork.com*
Street data for your town	• ESRI Data & Maps Media Kit • Data sites on the Internet (see Community Geography Web site) • U.S. Census Bureau TIGER/Line data from *www.geographynetwork.com*
Land-use or zoning data for your town (current and twenty-year projections)	• Your state's department of environmental management • Your town, county, or state department of planning • USGS National Land Cover Dataset from *landcover.usgs.gov* or *seamless.usgs.gov*
Public utilities data for your town (sewer pipes, water lines, gas lines, and so on)	• Municipal government offices • Specific utility companies servicing your community
Bodies of water (rivers, streams, watersheds, and so on) and flood zones	• U.S. Census Bureau TIGER/Line data from *www.geographynetwork.com* • USGS National Hydrography Dataset (NHD) from *nhd.usgs.gov* • Your state's department of environmental management • Flood zone maps from Federal Emergency Management Agency (FEMA) at *www.msc.fema.gov*
Elevation	• USGS National Elevation Dataset (NED) from *gisdata.usgs.gov* or *seamless.usgs.gov*
Aerial photographs	• Your town or state's GIS and planning departments • A GIS user in your town or region (utility companies, phone company, and so on) • Download digital orthophotographs (DOQs) from *www.usgs.gov* or *www.terraserver.microsoft.com*
Landmarks, homes, and structures	• Your town's GIS or public works department
Marked trails	• USGS paper topographic maps • Your state's department of wildlife • If trails are on federally owned land, check with the U.S. Forest Service, National Park Service, Bureau of Land Management, or U.S. Fish and Wildlife Service
Traffic flow and transportation data (use of area roads, public buses, trains, and so on)	• Your state or city's department of transportation • Your town or county's planning department

Tips on finding data about your community
- If your town's planning office uses GIS, they will be the ideal source of data about the characteristics of your community.

- If your own town does not use GIS, you may be able to locate appropriate data through your town's planning office, county or state government's office of economic development, or county or state government's office of regional or statewide planning.

- GIS users in your town or region (utility companies, phone company, and so forth) may also be willing to provide some of the data you need.

- Your state's Web site may identify sources of local data that are available to you.

Depending on the type of geographic question you've identified, you may need to collect data for your analysis. In the case study, the FLITE students collected data using GPS units. You may need to collect data on the location of unique features of the landscape (e.g., location of structures, signs, trees, overflow parking areas). If you need to collect data that is not already available, consider the following tips:

- Obtain enough GPS units for the data collectors.

- Establish a data collection protocol and train all data collectors in it.

- Practice data collection protocol with all data collectors.

- Set up an organized data table in advance of your collection. Be sure to iden-tify what data you need and how you plan to analyze it in the future. The type of analysis you plan to perform may influence how you record your data.

EXPLORE GEOGRAPHIC DATA

Technical issues
When performing a site selection, you will need to acquire data from a wide vari-ety of sources. One of the most important technical issues you will face is making sure each data layer can be viewed with all the others. Items to consider include:

- How important is the map projection? All of your data must be in the same map projection so that it can display in the same view extent. Quite often, aerial photographs come in a specific projection. If you obtain a digital photo-graph in a specific projection, make sure all the other data is in that same projection or you will not be able to see it correctly.

- Is the data lining up properly? Different map projections are a major cause of data misaligning with other data. Another reason could be differences in when the data sets were collected. For example, you may have an aerial photograph of a river taken in 2000 and a topographic map from 1960. In those forty years, it is likely that the river flow has changed and the two data sets do not line up. In this case, you may decide to use the current aerial photograph to digitize the location of the river. This is what the FLITE students did when they were presented with data from different time periods.

- Is there so much data that the map is difficult to view? An easy way to solve this problem is to set scale dependencies for each data layer. For example, you may not want to display the road network until the map is zoomed in to a particular extent. Set scale dependencies for different data layers so you can view those layers at an appropriate scale.

Preliminary data exploration
Before you're ready to analyze the data, you need to become familiar with it.

- Display all the data. Set scale dependencies (described above) and symbolize all of the data in relation to the criteria you have identified.

- Experiment with how the different data layers overlay. Reorder the themes in the table of contents so you can see all of the data.

- After looking at your data in map form, you may realize that you are missing critical information for your project. Locate any needed data and add it to the project.

- Incorporate photographs and map information to get a basic visual and spatial overview of the area.

ANALYZE GEOGRAPHIC INFORMATION

Determine your selection criteria before you begin analysis. For example, does the parking lot entrance need to be a minimum distance away from an intersection with another road? Are there preferred criteria that need to be identified and prioritized? Work with your community partner on ways to prioritize the criteria. Make a list of what your criteria are and start there.

- Plan the procedure you will use to select the site options. One method is to start eliminating all land that is not an option due to planning and zoning regulations. Another approach is to focus on identifying land that is most suitable for the site. You can begin with performing a query of elevation data to determine which areas naturally have the most appropriate slope and elevation for your site. Your town's planning and zoning regulations and the specific requirements of your site will help you determine where to begin analysis for your project.

- Eliminate or select site options by performing overlay analysis. For example, create buffers that reflect the laws and rules of the community about where the site needs to be placed. Perform theme-on-theme selection or query your data using the required and preferred criteria you identified earlier. With raster data, use digital elevation models (DEMs) to consider factors such as elevation, slope, or aspect.

- Draw a polygon that represents the size parking lot you need. Visually analyze possible locations for it. Determine general areas that are possible site locations.

- Select at least two possible sites as options. Review their benefits and disadvantages.

Tips for creating buffers

- You can create new shapefiles and draw your own buffers or zones. Besides being the simplest method, this method is also useful if planning and zoning guidelines use parameters other than distance (for example, the location of arterial streets or subdivisions).

- If you are using ArcView 3.0a, you can automatically create buffers of a specified distance using the sample extension *buffer.avx* that comes with the software. Instructions for running a similar tool are included in this module's exercise (pages 228–30).

- If you are using ArcView 3.1–3.3, you can create buffers automatically using the Create Buffers wizard. This tool will create buffers around a selected feature by using a wizard that walks you through the process step by step (see page 228).

ArcView extensions useful for site selection

SOFTWARE	WHAT YOU CAN DO WITH IT	WHERE TO GET IT
ArcView Spatial Analyst extension	Create, query, map, and analyze cell-based raster data to perform vector–raster analysis. Assign weights to multiple data layers and use Map Algebra to find cells that best meet your criteria for a new site.	*www.esri.com* or your local ESRI Business Partner. Educational pricing is available. PC only.
ArcView 3D Analyst extension	Create, analyze, and display surface data. Includes simple three-dimensional vector geometry that lets you perform line-of-sight analysis and create 3-D visibility maps useful in site analysis.	*www.esri.com* or your local ESRI Business Partner. Educational pricing is available. PC only.
ArcView Image Analysis extension	Use enhanced geographic imaging tools to quickly display and manipulate various image data. Create land-cover classes, clip and mosaic imagery, and perform change detection to supplement your site-selection analysis.	*www.esri.com* or your local ESRI Business Partner. Educational pricing is available. PC only.

ACT ON GEOGRAPHIC KNOWLEDGE

After analysis, it's time to put this valuable information to use. The information you have collected is vital to public officials for the planning and development of a parking lot or another site.

Possible action steps

- Create presentation layouts that visually display the entire area and your chosen sites. Use different maps to highlight the pros and cons of each possible site.

- Conduct presentations for your community partner and the public for decision making. Be sure to describe your process of site selection and why you chose each possible site.

- Write press releases announcing your presentations and invite local media.

- Give the data and project to your community partner so they have this information for future decision making.

- Educate others about the power of GIS to solve community problems by making public presentations or writing articles for local newspapers and magazines.

NEXT STEP

Completing a site selection is not the end of a project, but the beginning of a journey of community involvement and sustainable development. Follow up to evaluate the effectiveness of your selection. Continue to work with your community partner on projects related to site selection or other important community issues.

MODULE 8 ACKNOWLEDGMENTS

Thanks to the FLITE Team of 2002, their teachers, Colorado Division of Wildlife, and Connie Knapp of The Orton Family Foundation for sharing their community-mapping story with us.

Chuck Lewis State Wildlife Area boundary and feature data provided courtesy of the Colorado Division of Wildlife and is used with permission.

Routt County boundary, roads, and highway data provided courtesy of Routt County, Colorado, and is used herein with permission.

Flood-zone data provided courtesy of the Federal Emergency Management Agency and is used with permission.

Steamboat Springs data provided courtesy of the City of Steamboat Springs and is used with permission. This data was prepared from publicly available information and should be used for reference purposes only. Any other use or recompilation of this information is the sole responsibility of the user. This data cannot be used to establish legal title, boundary lines, setback compliance, locations of structures, improvements, or utilities, or relied upon in any flight activity. It will not be accepted as a substitute for ground site survey information during the planning/engineering process of project development. The City of Steamboat Springs expressly disclaims all liability regarding accuracy or completeness of this data.

On your own: Project planning
Building a community GIS project of your own

We hope that the seven case-study modules in the first part of this book have piqued your creativity and given you the confidence to begin your own community GIS project. Each community GIS project is grounded in the geographic inquiry process, the cooperation of multiple partner organizations, and the spirit of community service. At the same time, each project is unique—and yours will be, too.

This section presents information that applies to community GIS projects in general. It begins with some questions to help you define a framework for your project, and then provides information and tips for each step of the geographic inquiry process. You may find that a particular tip doesn't provide all the information you need. When you have questions, you may need to read up on a particular topic or consult with a knowledgeable community partner. (Suggested resources can be found in the references and resources section of this book. Additional resources can be found on the book's Web site, *www.esri.com/communitygeography.*) Use the material in this section in conjunction with the material found in the topic-specific "on your own" section of each module.

DEFINING A PROJECT FRAMEWORK

Once you have decided to do a community GIS project, how should you begin? You may find it helpful to start by asking yourself a few questions before you embark on the geographic inquiry process. Identifying the characteristics of your project up front will establish a framework for your project that will help you to quickly evaluate specific project ideas as they arise, and ultimately, choose a project that is appropriate for you. Your framework can be refined as the project develops.

Key project questions to ask
- What topics or community issues interest you or your group? At this point, you could identify a broad interest area, like "environmental issues" or "public health," or you could identify a more specific topic like "air pollution" or "bicycle trails."

- Why are you doing the project? What do you hope to accomplish? Identify the major goals of the project.

- Who in your community has knowledge about your broad topic of interest? Who in your community might want to be involved in the project? Make a preliminary list of possible project partners.

- How ambitious is your project? If this is your first project of this type, start small! If the project is successful, it can always be expanded later.

- What is the project time line? Do you need to complete the project by a specific date, such as the end of the school year? If your project might involve field data collection, would you need to complete your fieldwork before snow covers the ground in winter? How flexible is the time line? For example, are potential gaps in project activity desired or will they be a problem?

Once you have developed a framework for your project, you are ready to begin asking geographic questions.

ASK A GEOGRAPHIC QUESTION

In the first step of the geographic inquiry process, your goal is to identify one or more geographic questions related to a specific community issue.

Develop a geographic question to investigate

TASKS	SUGGESTIONS
Brainstorm ideas to identify issues in your community.	Ideas may be triggered in several ways: • Current events. Typically, something changes (e.g., the local newspaper runs an article on crime trends in the community, or the city unexpectedly acquires a piece of land for a community park). Consult local sources such as television, radio, community newsletters, newspapers, and Internet sources to identify current events. • A sense of place. This could be an awareness of your surroundings or observations you make about your community. What makes your community unique? For example, if your broad interest area is "environmental issues," are there factors related to climate, vegetation, wildlife, landforms, or natural-resource-related industries that make your community unique? • A community need arises. For example, the town builds a new elementary school and school administrators need to decide who will walk to the new school and who will take the bus.
Formulate geographic questions.	Once you have a topic or issue, identify its geographic components. Develop geographic questions that are relevant to the issue. For example, if your topic is reducing crime in your neighborhood, two geographic questions could include "Where is crime occurring?" and "Is there a pattern to when and where crime occurs?" If your topic is who will walk or ride the bus to school, you could ask "Where do the students live who will attend the new school? How is walking distance defined?"
Narrow the focus.	For example, if your community-mapping issue is "helping tourists visiting my town," you might narrow the focus to the historical sites that would be of interest. If your project is to identify potential sites for community gardens, you could focus on finding sites to serve a particular neighborhood.
Define the geographic extent of your research.	Does the issue affect only a small part of the community, or does it affect an entire county? Does the area it affects follow any obvious boundaries (e.g., rivers or streets)?
Develop one or more hypotheses about the problem or issue that you will test through your geographic inquiry and analysis.	Create "I think . . ." statements to help identify your knowledge and beliefs about the research question. Use this information to frame a working hypothesis. For example, if one of your statements is "I think crime has increased in the South Village neighborhood," your research could test whether crime actually has increased over the past year, or whether crime in this neighborhood has increased relative to other neighborhoods.

Set the stage for your project

- Select a study site. Consider the project framework you developed earlier, as well as factors such as extent, site accessibility, and community priorities.

- Perform research to understand the issue you select. When and how did the issue first arise? Are there conflicting points of view on the issue? If so, what are they? Are these represented by distinct constituencies in the community? What groups, if any, have been working on learning more or finding solutions?

- Identify what you will need in order to complete the project. For example, will you need GIS data that you don't have? Equipment to gather data, such as GPS units? Experts to consult? Will you need technical services, or will you provide them?

- Form partnerships with community organizations. Once you have a topic and questions in mind, you are ready to reach out to potential community partners. These include local government, businesses, nongovernment organizations, community groups, and schools. Use your topic of inquiry, needs, and background research as your guides. If an organization you approach is not interested in partnering with you, ask for names of other organizations that might be interested. Also, think about how the schedules of school partners and other community partners can fit together.

- Write a research plan that begins with a statement of the issue or problem to be studied and briefly outlines the basic plan for the project.

ACQUIRE GEOGRAPHIC RESOURCES

When you have settled on a specific community issue and identified the pertinent geographic questions, the next step is to acquire the geographic data and other resources you will need to map and analyze.

To create maps using a GIS, you need the right data. Data for a GIS comes in three basic forms:

DATA	EXAMPLES
Map data is made up of points, lines, areas, and rasters. Map data forms the locations and shapes of map features such as buildings, streets, and cities. Map data can also show natural and human-made landscape patterns such as rainfall, temperature, soil type, zoning code, or population density.	shapefile (.shp, .dbf, .shx), ArcInfo™ coverage (may be in compressed format, .E00), geodatabase, ArcGrid™
Tabular data is information describing a map feature. For example, a map of customer locations may be linked to demographic information about those customers.	text file (.txt), dBASE® file (.dbf)
Image data includes such diverse elements as satellite images, aerial photographs, and scanned data—data that has been converted from paper to digital format.	JPEG (.jpg, .jpw), geotiff (.tif, .tfw), ERDAS IMAGINE® image (.img)

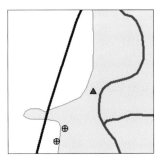

Map data

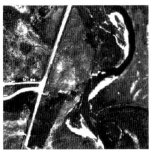

Image data

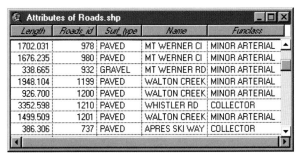

Tabular data

Identify the data you need

As you search for data, you may know exactly what data sets you want and where to get them. It's just as likely, however, that you will go through a process of making a wish list of data set characteristics and then looking for data sets that meet your criteria. At this time, begin thinking about the analysis you plan to do and decide on a preliminary problem-solving approach. What you want to do with the data in the GIS will determine the appropriate form of data (map, tabular, or image) as well as the characteristics of the data. For example, if you need streets, a data set that is fine for drawing a street map may not include the information necessary for geocoding or for developing bus routes. Or, if you want to show streets as an underlying theme with other data layers, a digital orthophotograph could be the best choice.

The following list covers the most important issues you will need to consider, either by yourself or with the help of a knowledgeable GIS expert, when identifying the data you need.

DATA ISSUE	CONSIDERATIONS
What are the specific geographic features you need?	If your project involves streets, do you also need bridges, traffic signs, crosswalks, or rivers? What data do you need to provide context for the issue of concern?
How much detail about those features do you need?	To gain the most understanding from your GIS, determine the level of detail required. For example, do you need all streets or major highways at a "local" scale (1:24,000), or major highways at a "national" scale (1:3,000,000)? Even for a specific feature like streets, you may need to decide how you want them represented (centerlines, double-lined streets, or connected routes).
What attributes of those features do you need?	The attributes you need depend on the task at hand. Street data to be used for geocoding, for example, needs to have attributes such as street name and address ranges. If you want to calculate the number of miles of each road type found in your county, you'll need attributes like road type (highway, arterial, local) or surface class (paved, gravel).
What is the geographic extent of your interest?	Are you examining a neighborhood, a metropolitan area, a state or province, or an entire country? Factors that could affect how much data you acquire include availability, cost, and storage size (especially for image data).
What is the level of geography you want to examine with your area of interest?	If you are using demographic data, for example, what is the summary level of geography you want to examine? Are you examining statistics across your state by census tract, block group, ZIP Code, or school districts? If you have a table of statistics by census tract, for example, you will need census tract boundary data to map the statistics.
Is currentness of the data important?	For some applications, such as land-use planning using remotely sensed imagery or aerial photography, obtaining current data will be a priority. If you are studying changes in a community, it might be important to obtain data for multiple time periods.
What computing environment will you be using? Do you need to request data in a preferred format?	For example, will you be using Windows 2000, Windows NT®, Macintosh, or a UNIX® workstation? The answer to this question can affect which software products you will be able to use, which in turn may affect the data formats you need to acquire (e.g., shapefile, text, dBASE).
What map projection and coordinate system does the data need to be in?	In most cases, all the data sets you are using should be in the same map projection and coordinate system. You may want to decide on an appropriate map projection and coordinate system for your project so you can specify this information when requesting data. Or, you may acquire a key data layer or set of layers and choose to adopt whatever map projection they have.
When do you need the data?	Many data sets can be downloaded immediately from the Internet. Others may take weeks to request, prepare, and be delivered by a data provider. Depending on your project schedule, you may need to choose a data source based on how quickly you can get the data.
Do you want to publish maps or other derivative products with the data?	Do you foresee creating hard-copy maps that you will give away or sell? Do you plan to publish the data on an intranet or the Internet? Many data publishers restrict such redistribution. Your best bet is to check with the data provider to see if your intended use is permitted.

Pointers on where you can get data

Many data sets are available for free or at low cost from government agencies and data clearinghouses on the Internet. Data clearinghouses are basically huge lists of other sites that have data (free and for a fee). Consult this table for a quick reference on where to look for a particular type of data.

WHERE TO LOOK FOR DATA	TYPE OF DATA
Your local town, county, or state GIS department	• Street data for your town or state • Local waterways • Landmark or historical site locations • Land-use data • Parcel and zoning data • Major transportation routes (train, bus) • Aerial photographs of your town
Your local town, county, or state specialty departments (e.g., police, public works, parks and recreation)	• Specialty data such as crime, police patrol zones or precinct areas, fire hydrants, drainage ditches, parklands, urban trails, and so on
The U.S. Census Bureau (*www.census.gov*)	• Demographic data such as income, race, age, employment category, sex, level of education • Populations of towns, cities, states
Geography Network (*www.geographynetwork.com*)	• Basemap layers and other worldwide data from many data providers; census demographic information and street data
USGS (*www.usgs.gov*)	• Topographic maps of the United States, elevation data, land-use or land-cover data, hydrologic (water) data, biological (botany, fisheries, reptiles, biodiversity) data, earthquake data, digital raster graphics, orthophoto quadrangles, and a variety of other data sets
Government agencies, such as fish and wildlife or environmental protection, for your state	• Localized data on the environment, habitat, locations of wildlife, and presence of chemicals or pollutants

Collecting and creating your own data

In many GIS projects, you will have to collect or create some of the data. When planning how to collect your data, you should take the following points into consideration:

- *Acquire necessary equipment.* If you need GPS units, water-quality testing kits, measuring tape, or other collection devices, you must research the different products and determine which ones will fit your need. For example, your project may not require a high-precision GPS unit—an ordinary model may be adequate.

In the Barrington tree inventory (module 7), students used the equipment pictured here to collect their data.

- *Identify what data you need to collect.* Create a spreadsheet so you know which attribute information needs to be collected.

- *Acquire data by questionnaire or interview.* If you will be collecting data by questionnaire or interview, use your spreadsheet (above) as a guide for preparing the questions you will ask people. After you have collected the data, tabulate the results and enter them into a spreadsheet that you can display or map in ArcView.

- *Determine your data-coding method.* Some data is better represented in a database by using a code. For example, using the code "RM" is more efficient and less error-prone than typing "red maple" into the database. In addition, the type of field you select for each attribute (string, number, Boolean, or date) will determine the type of analysis you can perform. In ArcView, for example, statistics can only be calculated on number fields. You cannot calculate statistics on numbers stored in text string fields. Determine how many decimal places your numeric data needs to have.

- *Give each observation (record) a unique identification number or code.* Unique IDs are an essential component of GIS database design because they provide a reliable way to assign attributes (your observations) to the correct features. Unique IDs provide other advantages as well. For example, when you enter data, it is easier to check for data errors if you know which record corresponds with which feature. Unique IDs make it possible to join your data table with other tables containing information about the same locations. Also, you need to decide how to handle situations where a feature is made up of multiple shapes. For example, Hawaii is made up of several islands. Should each island have a different ID, should they all have the same ID, or both? If one record in your table applies to the state as a whole (e.g., state population), giving all of the islands the same ID will allow you to correctly calculate statistics like total U.S. population.

- *Establish data-collection protocols.* In order to collect good data, you should establish collection protocols according to industry standards and educate data collectors about these important policies. For example, if five different people are collecting water-quality data, it's important that each follows the same procedure so you can compare the data.

- *Practice collecting your data.* Before collecting your data, practice collection techniques and recording the data into the database. At this point, you may find that a procedure needs to be changed or a recording method altered. Ironing out these details before you begin the actual collection will help the project in the long run.

- *Perform quality control after data has been entered.* Errors in data entry will result in errors in data visualization and analysis. Finding input errors now will avoid problems later on. Make sure the database is complete (no missing values). Review data for logical consistency. For example, check to see that values are within a reasonable range (a tree should not have a diameter of 10 feet).

- *Create documentation about the data (metadata).* Other people may be interested in your data, either to understand more about your project or to use it in their own research. However, what makes perfect sense to you may not be so clear to others. Make a record of what the field names and codes mean. Describe your methodology. Include the date the data was collected, the map projection, if applicable, and other standard pieces of metadata. Refer to the data permissions and use restrictions section below for more information about metadata.

In some GIS projects, you may need to create new data from existing data using the GIS software. In ArcView, there are two basic ways you can create data.

1 *Derive data from existing geographic themes.* You can do this by performing analysis techniques such as queries and summaries to form new shapefiles, views, or charts.

2 *Digitize paper maps or imagery.* You can trace shapes such as a building outline from a paper map to create new shapefiles, or you can scan a map or aerial photo to create new image layers. You can also perform on-screen digitizing, also known as *head's-up digitizing,* by tracing shapes from an image (such as the outline of a building) to create a new shapefile.

Developing and managing your database

Whatever the size of your project, it is important that you develop a plan for where and how you are going to store all your data. Typically, you will have several versions of your database. These include original or raw data (e.g., tabular data in a spreadsheet, data downloaded from a GPS unit, data from a community partner), data that you have modified or created (e.g., ArcView projects, event theme shapefiles, projected data), data resulting from your analysis (buffers, query results), and data created to support your actions (layouts, presentation slides, handouts). You may need to decide how to handle copies of data for multiple users, as well (e.g., multiple ArcView projects, multiple analysis result files). Developing a good management plan in the beginning will save many headaches in the end. Here are some items to consider:

- *Develop a consistent directory structure.* Decide how you want to organize your data and keep your method consistent throughout.

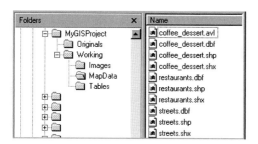

One way to organize your data is to create separate folders for original and working copies of the data. In this example, the data is organized by type within the Working folder. Other ways you could organize the data are by extent (e.g., county, city, study area), location (city A, city B), type (basemap, project data), date (historical, current), or user (student A, student B).

- *Establish a file-naming convention.* If you will have multiple people saving and editing data, having a naming convention is essential. In this book, we encourage people to save their project files with their initials at the end of the file name. You may decide to include as part of your convention the date the data was saved. No matter what naming convention you decide to use, make sure everyone working on the project understands it and uses it.

RANDOM FILE NAMES	FILES NAMED USING A CONVENTION
50buf.shp	river500buf.shp
stream200.shp	river200buf.shp
buffer3.shp	road50buf.shp

A file-naming convention can help you organize your data and find files when you need them. If you used the file names at the left, for example, you might not remember which file represents a road buffer. If you use "streams" for a buffer of river.shp, the two files may not be near each other in an alphabetical file list. Often, you can include several elements in the name; the names at the right include the type of feature, the buffer distance, and the fact that it is a buffer.

- *Decide how you will manage different "versions" of the data.* At the least, you should have two versions of your data—a master database and a backup copy. If you will be creating or modifying data (e.g., downloading data from GPS units, adding a field to a table, editing features in a shapefile), it is wise to keep a copy of the original data in case you make a mistake or something goes wrong. Different versions also will exist when many people are working on separate portions of the data that will eventually be combined.

- *Determine where the master data and versions will be stored.* Will your data be stored on a server or on a local machine? Who will have access to the data? Decide where to store the master copy of the data and where to store interim versions (including original hard copies like data-collection sheets or surveys, if applicable).

On your own: Project planning 249

- *Designate a location for supporting documents.* Supporting documents include a data dictionary that explains all of the attributes, any "readme" files, or research that has been conducted. These documents will be important to the people accessing your data and should be made available at a central location (such as a server) accessible by everyone working on the project.

Data permissions and use restrictions

Most of the data that you collect or use in your GIS project will have metadata associated with it. Metadata is frequently described as "data about data." It is additional information about the data that helps a user understand the data attributes, geographic area and scale, who created the data, and restrictions on its use.

When working on a GIS project, it is important to be aware of data license agreements or any use restrictions for the data you are using. While gathering data, you need to make sure you have permission for your intended use. Many data providers place restrictions on redistribution of data (e.g., they do not allow the data to be sold or copied). In some cases, data providers will grant you special permission to redistribute data. Contact the data provider, explain the nature of your project, and ask the data provider for permission. Expect to compensate the data provider with credit in your published work and in some cases, with money.

In addition, you need to create metadata for the new data you have created. You may want to restrict the uses on your data and will need to include that information within the metadata documents.

For more information on metadata and use restrictions, visit the FGDC (Federal Geographic Data Committee) Web site at *www.fgdc.gov.*

This metadata for aquifers.shp (module 4) is displayed as a set of frequently anticipated questions. You can view metadata in a Web browser or with a text editor program.

aquifers.shp

Metadata also available as

Frequently-anticipated questions:

- What does this data set describe?
  1. How should this data set be cited?
  2. What geographic area does the data set cover?
  3. What does it look like?
  4. Does the data set describe conditions during a particular time period?
  5. What is the general form of this data set?
  6. How does the data set represent geographic features?
  7. How does the data set describe geographic features?
- Who produced the data set?
  1. Who are the originators of the data set?
  2. Who also c
  3. To whom sh
- Why was the data
- How was the data
  1. From what
  2. How were t
- How reliable are th
  1. How well ha
  2. How accura
  3. How accura
  4. Where are t
  5. How consis
- How can someone
  1. Are there le
  2. Who distrib
  3. What's the
  4. What legal
  5. How can I
- Who wrote the me

What does this data set describe?

Title: aquifers.shp
Abstract: Principal and Glacial Drift Acquifers of North Dakota

1. **How should this data set be cited?**

   ESRI, Unknown, aquifers.shp: Environmental Systems Research Institute, Inc. (ESRI), Redlands, California, USA.

   Online Links:
   o \ESRI\ComGeo\module4\aquifers.shp

2. **What geographic area does the data set cover?**

   West_Bounding_Coordinate: -97.859741
   East_Bounding_Coordinate: -95.379318
   North_Bounding_Coordinate: 49.000023
   South_Bounding_Coordinate: 46.646366

3. **What does it look like?**

Privacy and data

Depending on the subject of your GIS project, you may need to consider privacy issues, such as a data provider's need to maintain the privacy of individuals. For example, if you are trying to determine which elementary students walk to school and which take the bus, the identity of individual students might be withheld. Most likely, you will be given a database that has randomly assigned student numbers, addresses, and ages, while names, social security numbers, and personal information are withheld. In fact, you might ask that such personal information be omitted when requesting the data, to avoid any problems later on.

When you work with data that has been modified, it is important to know how it has been modified. You should always ask for this information from the data provider. In some cases, the modification may affect your intended analysis. For example, if you are doing a crime-pattern analysis and the police department eliminated all sex crimes, you will not be able to do an analysis that accurately shows the percentages of different types of crime.

The Freedom of Information Act and data

In the United States, you can obtain many kinds of data through the Open Records provision of the Freedom of Information Act. This federal statute applies to all government agencies within the United States including local law enforcement agencies. To request data through the Open Records provision, you must submit your request in writing either through standard mail or fax (e-mail requests are not accepted). Send your request to the government agency that has the data you want. According to the law, agencies have twenty working days to provide an initial response. The agency may contact you for clarification to ensure they provide the correct data. If the agency does not have the requested data in electronic format, which is often the case in smaller towns, they will provide it in written form and you will need to enter it into the computer. Usually a small handling fee is charged to cover the cost of materials and shipping.

The Freedom of Information Act Web site is *www.foia.state.gov.*

EXPLORE GEOGRAPHIC DATA

Once you have acquired data, you will need to display it in the GIS and then explore it to become familiar with it.

Preparing data for exploration

Quite often, preparing the data for display can take significant time. For example, when the John McCrae Secondary School students in Ottawa, Canada, (module 6) wanted to display the student data they received from the elementary school principal, they spent a lot of time geocoding it. If your project involves displaying address location data, you will need to geocode the data before you will see it as points on your map.

- If you have coordinate locations (latitude/longitude, northing/easting, and so on) that were downloaded from a GPS unit, you may need to add that data as an event theme to display it in ArcView. Each GPS unit is different in how it records data, so you will have to check with the manufacturer of your brand for details. (Refer to the module 3 and 4 exercises for information on how to create an event theme from GPS data.)

- You may obtain data that is projected differently from other data. In some cases, such as when you have an image that is projected and other vector themes that are unprojected, you can still display them together in ArcView by changing the projection setting in the View properties dialog to match the image's projection. Another option if you have ArcView 3.2 or 3.3 is to use the Projection Utility Wizard extension. This extension lets you project shapefiles from one coordinate system to another.

- If you have image data that is not appearing in the same geographic location as your other themes, the image may need to be georeferenced. If you are working with a JPEG (.jpg) or TIFF (.tif) file, check to make sure the world file (the file that contains the geographic coordinates) is included and is named correctly (.jgw or .tfw).

- When you are symbolizing data, consult with community partners for guidance on how to symbolize the data in ways that will be meaningful to the intended audience. They can provide information on colors, symbols, and classification methods commonly used to display a particular type of data.

Exploring the data

Usually the best way to begin exploring your data is to create a basemap of your area of interest. A basemap typically depicts geographic features such as landforms, drainage, roads, landmarks, and political boundaries. Sometimes it is appropriate to create one basemap at a small scale (e.g., a state or county) and one at a larger scale (your project area). These could be two different views, or one view with scale-dependent themes (see the module 8 exercise for an example).

Explore your basemap to understand the geographic context of your project. For example, if you are looking at an urban area, where is the street network most dense? Where is the downtown district? Is the city located near a body of water?

After constructing and exploring your basemap, you can add more data to your map. The basemap themes will provide locational reference for the other themes (robberies, aquifers, tree locations, buffers, and so forth).

Tips on displaying your data

TIP	WHAT TO DO
Classify the data in a variety of ways	Vary the classification attribute, classification method, number of classes, and so on. Observe how your choices affect the appearance of particular geographic patterns or relationships. Could the differences you observe result in different conclusions about the same data? (Refer to Monmonier, *How to Lie with Maps*.)
Set scale dependencies	Set minimum or maximum display scales so themes are displayed only at appropriate map scales as you zoom in or out.
Turn layers on and off	Explore the geographic relationships among different combinations of layers.
Experiment with colors and symbology	Find colors that provide contrast with other data classes or other themes. Add topic-specific symbol sets to the symbol palette (see the module 2 exercise). Decide whether graduated symbols or graduated colors are more effective for portraying the geographic distribution of your data. See what symbol size or line thickness works best. To get ideas from professionally produced maps, browse the Map Gallery on ESRI's Web site *(www.esri.com)*.
Reorganize the data into themes and views	Decide what data to show together. If you have many themes, use multiple views to make the data display more manageable. If one shapefile contains many attributes or features, use multiple themes to display different attributes or perform queries to display subsets of the features.
Modify or add text labels	Use descriptive names for views, themes, and data classes. Add text labels to key map features.

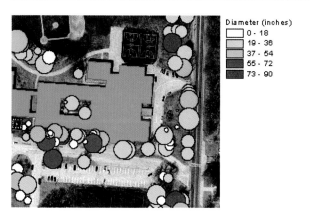

Diameter (inches)
- 0 - 18
- 19 - 36
- 37 - 54
- 55 - 72
- 73 - 90

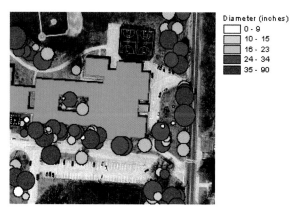

Diameter (inches)
- 0 - 9
- 10 - 15
- 16 - 23
- 24 - 34
- 35 - 90

This map of Barrington Middle School trees uses five equal-interval classes. The range of diameter values from smallest (less than 1 inch) to largest (90 inches) is divided by 5 so that each class has a range of 18 inches (0–18, 19–36, and so on). You could conclude from the map that most of the trees have a small diameter, and that those with a large diameter tend to be clustered together.

This map uses quantile classification. Here, the total number of trees (217) is divided into five classes of approximately forty-three trees each. Even though this map uses exactly the same attribute values as the map at the left, you would probably draw a different conclusion. This map implies that there are many large trees and that they are fairly evenly distributed around the school.

While displaying your data, spend time identifying geographic patterns and relationships. Examine which attributes are present and the range of their values. Review your geographic questions and make sure the attributes you now have will be helpful in answering those questions. Look for visual patterns and relationships. Write everything down and refer to these notes later as you analyze the data. Consult with community partners for feedback and support during this process. They can help you identify and interpret geographic patterns and suggest ways you can prepare for the next step in the process: analysis.

ANALYZE GEOGRAPHIC INFORMATION

Analyzing geographic information is all about turning the pieces of data you have acquired into geographic knowledge. This step incorporates three activities: (1) deciding what problem-solving approach and specific steps you will use, (2) performing the actual analysis using the GIS or other tools, and (3) drawing conclusions. Depending on the complexity of your central geographic question, you may need to do these activities more than once before your analysis is complete.

Choosing an analysis approach

In general, GIS analysis is a process for recognizing geographic patterns or relationships between geographic features in your data. Remember that almost always there are multiple ways of getting the information you need from the GIS. Often, one approach is quicker and gives you more approximate information. Others may require more detailed data and more processing time and effort, or more specialized software tools such as ArcView extensions. For example, if you were analyzing change over time, you could create a series of maps to show snapshots for two or more times. You would visually compare the maps to see where and how much change occurred. Alternatively, you could use the GIS to measure the change, and map the calculated difference between two time periods. This approach would likely take more effort but yield more precise results.

The approach you use depends on the following factors:

- The type of problem and how deeply you understand it
- The kind data you have
- The quality of your data
- The GIS tools you have available and your fluency with them
- How much time you have
- How defensible your methods need to be

Tips on choosing an analysis approach

- Keep in mind your geographic question when determining what kind of analysis you will do. Ask the question (or questions) as specifically as possible. In module 5, for example, Crescent School students turned the general question "What landfill sites may pose a threat?" into the specific question "What landfill sites are within 0.5 kilometer of a school or river and in a residential area?"

- Think about how you can use the GIS to help you answer the question. Create a flowchart to help you develop your analysis approach. You cannot anticipate every problem you will encounter, but you should have a general understanding of the steps you need to take and which GIS operations support these steps.

- When developing your flowchart, begin with a simple outline and then fill in details like what GIS functions you will use (e.g., query, buffer) and their inputs and outputs. For example, what shapefiles or themes will be needed for each step? Are there tables you will need to join? Maybe you need attributes associated with points, but you have polygons, so you will need to extract centroids. Note on your flowchart any new files or themes that will be created. You could include specific information like query expressions or new file names, too.

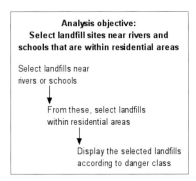

Create a flowchart when planning your analysis approach. At first, use descriptive language to outline the steps you will use to answer your geographic question. This example outlines the Analysis portion of the module 5 exercise (part 3, steps 1–10).

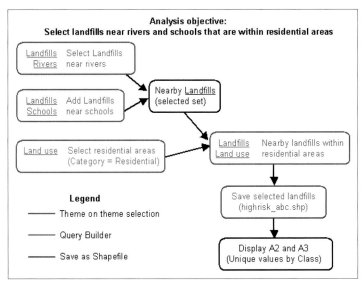

Next, work out how you will perform the steps using the GIS. You could use colored markers and paper or a whiteboard to create a flowchart similar to this one. Don't be afraid to refer to ArcView or ArcView help to confirm how a particular function works as you add the details.

- Keep in mind the purpose of your analysis. How will the analysis results be used? Who will use them? If you seek a better understanding of a place, a simple approach like visual analysis may be adequate. Consider the module 6 school busing study, for example. In answering the question "Where do the students live?" you might observe on the map two large clusters of students— one near John Young Elementary School and another in a neighborhood to the east. If your goal is to inform public discussion or decision making, you may need to be able to justify or describe your methods. The elementary school principal needed to know specifically how many students live within specific distances from the school, and she needed to know what technique was used to determine the distance.

- Consider what kind of information the attribute or cell values represent (e.g., categories, ranks, counts, amounts, ratios). A given analysis approach is typically appropriate for only some kinds of information. For instance, if you have tree data with height attribute values of short, medium, and tall (ranks), you will not be able to calculate the average tree height (an amount).

- Consider the quality of your data, including its accuracy and precision. A common mistake is to produce results whose accuracy or precision is not supported by the accuracy or precision of the input data.

Know the accuracy of your data so you can properly interpret your analysis results. Imagine that you used GPS to collect weed locations (blue triangles) that are accurate within 3 feet. You want to find all the locations within 50 feet of the road. If the positional accuracy of your road data is 30 feet (i.e., a road on the map could appear up to 30 feet away from its true location), then your analysis results will likely include some weed points that are not actually within 50 feet of a road (like point C) and miss others that are (point A).

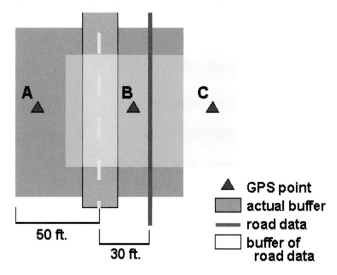

- The book *ESRI Guide to GIS Analysis, Volume 1: Geographic Patterns and Relationships* provides an excellent conceptual overview of many geographic analyses that you could perform with GIS. Refer to the reference section for more information on this and similar resources.

- Consult with your community partners. Ask them to share their subject-area expertise that may help you decide on an appropriate approach to the analysis, or ensure that you are using a widely accepted analysis method.

Performing GIS analysis

In the previous section, "Explore geographic data," you familiarized yourself with the data and performed preliminary steps such as geocoding or adding GPS data to the ArcView project. Once you have determined the analysis approach you will use, it is time to add extensions or scripts that you will be using for your analysis. Depending on the complexity of your analysis, you may want to develop a flowchart, create a backup database, and practice using the GIS tools. Once you are underway with your actual analysis, do not be surprised if you discover that you need to modify your analysis approach, acquire different data, or make other adjustments.

Tips on performing GIS analysis

- Refer to your flowchart. Using it as a checklist could be especially helpful if your analysis will take much time.

- Revisit your plan for naming and storing any new data files that will be created from your analysis. Do you need additional naming conventions? If necessary, set up a folder or folders for the new files. New files could include ArcView projects, tables, shapefiles, image subsets, and so on. If you will be editing data (e.g., adding an attribute field or new features to a shapefile), it is a good idea to save a copy of the original shapefile first, as a backup.

- You may want to practice with the analysis tools you plan to use before you do your actual analysis. If you need to, read the ArcView help to understand any input parameters or options the software presents you with. Make sure you know what form of output a particular tool produces (e.g., shapefile, graphic, temporary table).

- If your analysis will be complex, you may want to test some of your analysis steps using the actual data to be sure they will work the way you expect. If you will be analyzing a large amount of data, consider testing the methodology using a small subset of the data as a prototype. Even though you may be anxious to get on with your analysis, such tests could save you significant time in the long run.

- Take the time to visually analyze the data produced by intermediate steps as well as the final analysis. Are the results what you expected? Did the process you performed appear to have worked correctly? What new geographic patterns and relationships do you see?

Drawing conclusions

It is important to draw conclusions from your analysis. For example, if you are analyzing tree data (as in module 7) and you find that all the maple trees have a large diameter, you may conclude that they are all older trees. What inferences can you make to explain your observations? If the diameters of the maples are similar, you may infer that they were planted at the same time. This may lead you to further conclude that some young maple trees ought to be planted.

Tips on drawing conclusions

- Display the results in a variety of ways to see what patterns stand out.
  - As appropriate, try displaying the analysis results in different formats: maps, tables, charts, and so on.
  - Try displaying the data using different classification methods (e.g., natural breaks, quantile). Remember that the default classification method is often not the best choice.
  - Try displaying the results with a variety of other layers. For example, do the results of a crime study tell a different story when displayed with a roads theme than when displayed with a demographic theme of household income?
- Look at the results of your analysis to decide whether the information is valid or useful. Consider whether the analysis needs to be altered or refined.
- Ask your community partners to review your results. Their familiarity with the study area or topic of study may provide you with some additional insights.
- Remember to relate your conclusions to your geographic question or questions. If new questions come up, you may want to perform additional analyses before moving on to the next step of acting on geographic knowledge.
- Expand your thinking beyond the GIS analysis. Think about factors related to your issue or problem that could be relevant causes or effects of the community issue, but that you did not or cannot analyze with the GIS. Include these factors in your final recommendations and observations.

ACT ON GEOGRAPHIC KNOWLEDGE

Acting on geographic knowledge is a step that is often overlooked in the geographic inquiry process, yet it is vital if you want your project to make a difference in your community. You cannot just print your map and hope that someone else will notice—you need to go out and do something yourself! Figuring out what to do and actually doing it typically involves a process that can be summarized in these five steps:

1 Identify actions

2 Develop your message

3 Create visuals

4 Articulate the story

5 Follow up and evaluate

Identify actions

Once you have drawn your conclusions, you are ready to identify the action steps you can take. Before you identify the steps, review the original geographic question, the audience you are trying to address, your goals for the actions you take, and your options. You and your community partners may decide that different actions are appropriate for different audiences. You may also decide that you have time for some actions, but not enough time to prepare others. Moreover, you may find that a given objective can be accomplished through more than one type of action—consider which types of actions will be effective given the resources available to you.

For example, if your goal is to educate the community about an issue, you could (a) mail a report about your project to everyone in your community, (b) create a Web site, or (c) make a series of public presentations. A mailed report will ensure that everyone receives the information, but it may be costly to print and mail. Creating a Web site could cost less than a mailing, but a smaller group of people will be reached (people have to know about the Web site and choose to visit it). Public presentations would provide opportunities for feedback or discussion but may take more careful preparation.

POSSIBLE ACTIONS

- Issue a press release.
- Perform interviews for local newspaper or television reporters (do not overlook community cable-TV channels).
- Develop a Web site (or publish information on an existing Web site).
- Present information, results, or recommendations to a decision-making body (e.g., town council, school board, police department).
- Provide hard-copy or digital maps or data to appropriate community partners.
- Create educational materials (e.g., maps, brochures, posters).
- Perform community service (e.g., planting trees, cleaning up trash).
- Organize a public forum, rally, or protest.
- Present methods and findings to other schools or professional organizations that may want to use your project as a model for other communities.
- Continue work on the community issue.
- Teach community partners how to view or analyze data with GIS software.

Once you have selected an action (or series of actions) to take, create a plan outlining a time line for your action(s) and the resources required. For example, if you are developing a press release, who on your team will write it? Who will edit it? To what organizations will you send it? If you will be making public presentations, your time line should include time for preparation (creating maps and slide shows, reserving a meeting hall, publicity, and so forth), practice, presentation dates, and follow-up (creating reports of discussion, sending thank-you letters, refining the presentation).

Develop your message
Now that you have selected a course of action, think about what you want to tell your community about your efforts. You should be able to state, in a few concise sentences, the problem or geographic question you studied, your findings and conclusions, your proposed action, and the expected benefits. When you develop the message you want to communicate in your action, be sure to consider your audience and the type of information you have. Consult your community partners for their ideas, too.

Once you have developed your message or messages, you are ready to create visual materials to help communicate that message to your audience(s). What is the best way to communicate your message? A map might be the most effective way to present some types of information, while a chart or spreadsheet could be better in other cases. The Barrington eighth graders (module 7) chose to create a chart to display species diversity and a map to display the location and number of hazardous trees. Some schools have TV production facilities that students could use, for example, to videotape a multimedia presentation.

Create visuals
Visuals like posters and maps enhance presentations and will help to convey your message. Budget enough time to prepare them. In some cases, you may need to prepare a variety of visuals. For example, the Shelley High students in Idaho (module 3) created a wide variety of visuals to suit each audience. They created a map for the local and state weed officials, a weed identification guide in layman's terms for the general public, and a coloring book for elementary school children. You should also consider making visuals that can be reused. For example, you could create a map that becomes a poster and is also used in a slide-show presentation.

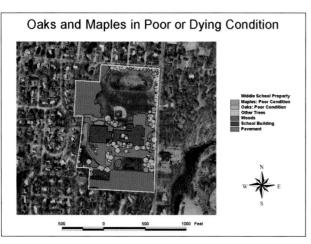

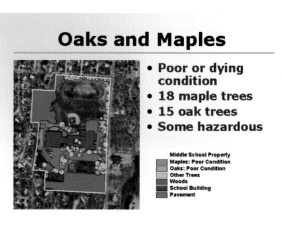

The layout at the left and the presentation slide at the right both use the same map created in ArcView.

Here are some things to consider when you develop your visuals:

- *Cartographic standards.* When you are making a map, choose your colors, labels, and symbology according to common cartographic practices. Also, consider the story you want the map to tell and the audience. If your audience is a group of elementary school students, you may want to use large text and easy-to-recognize symbols. If your audience is a group of adults who are experts in their field, you may want to use the industry standards for symbology. Do not forget to include a title, date, author, legend, orientation, and scale on all maps.

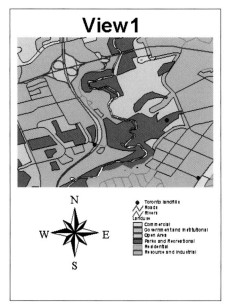

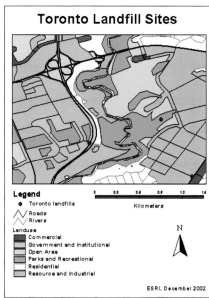

The layout at the left is difficult to understand because of poor color choices (red river, low-contrast land-use colors), missing information (lack of title and scale bar), and unbalanced graphic elements (large compass rose, small legend). The right-hand layout is more intuitive (blue river, green parks and open space), includes all essential information, and elements have a pleasing and functional arrangement.

- *Graphic design.* When preparing posters for public display or educational materials for distribution, good graphic design is essential. It could make the difference between someone reading and using your materials and someone misunderstanding or ignoring them. Make sure your fonts are legible and appropriately sized. Use a simple design that is easy to read.

- *Printing resources.* Think about how you are going to reproduce your visuals before you invest a lot of time creating them. Large-format graphics are normally output on expensive plotters. Perhaps one of your community partners has one of these and will let you use it to output posters and large maps. If you are printing color brochures or handouts, try to use your organization's resources or those of a community partner before paying a print shop to do the work. If necessary, include printing resources in your requests for funding.

Articulate the story

Tell other people about your results! You can generate a lot of public interest and potential funding for future projects if people know about the value of your work. There are many ways to inform the community. Use the communication channels (newsletter, magazine, Web site, monthly meetings, and so on) of your organization and your community partners. Here are a few suggestions on working with the media:

- Write a press release that addresses the "who, what, where, when, why, and how" of your project.
- Distribute the press release to local newspapers and television stations.
- Invite newspapers and television stations to a presentation or on-site demonstration.
- Prepare for media interviews by designating people to be interviewed and preparing something to show the reporters.

ORAL PRESENTATION TIPS

Oral presentations are usually an important action step. In addition to budgeting the time needed to prepare such presentations, we have other tips:

- Practice! Practice many times in front of people. Know the material well enough so you can consult your notes rather than reading them.
- Time your presentation. Make sure it fits within the time allotted.
- Know your audience.
- Tailor the presentation to how your results can benefit your audience.
- Make sure your presentation has a beginning, middle, and end.
- Account for technical needs. Bring your own computer, projector, screen, and so forth.
- Allow for adequate time to set up your equipment.
- Have water on hand during the presentation.
- Dress appropriately.
- Make sure people know about the presentation; you want them there to hear it.
- Have a plan B. What happens if your presentation time changes or if the equipment they promised is not there?
- Be prepared for the unexpected.

Follow up

After you have made your presentation or taken other action steps, some follow-up steps may be necessary. For example, you may need to write up notes on the public discussion at your presentation to include in your final project report, write thank-you letters, or talk to newspaper reporters about your action experiences. If you will be repeating your presentation, you might want to refine your presentation materials as a result of questions or comments from the first time.

EVALUATION AND NEXT STEPS

Once you have completed your actions and follow-up, you are ready to evaluate your project and think about possible next steps. How successful was your project? Did you make recommendations that were used? Did they work? Has this project spawned ideas for other related projects? If it has, then you should develop plans with community partners to begin work on a new project. Remember—the geographic inquiry process is iterative. Quite often, acting on geographic knowledge will mean asking a new geographic question. Then you can begin the process all over again!

References and resources

GENERAL RESOURCES

Convis, Charles L. Jr., ed. 2001. *Conservation geography: Case studies in GIS, computer mapping, and activism.* Redlands, Calif.: ESRI Press.

Diresta, Diane. 1998. *Knockout presentations: How to deliver your message with power, punch, and pizzazz.* Worcester, Mass.: Chandler House Press.

Envirolink: The Online Environmental Community. Resources on many environmental topics. www.envirolink.org

ESRI. Information about GIS software, data downloads, and applications. www.esri.com

ESRI. Understanding Geographic Data. ESRI Virtual Campus course. campus.esri.com

Federal Geographic Data Committee. An excellent resource for spatial data and information on metadata. www.fgdc.gov

Freedom of Information Act. Provides information on how to use this act to obtain data. www.foia.state.gov

Geography Network. Comprehensive source for online GIS maps, data such as roads, U.S. Census 2000 TIGER/Line data, demographic data, and services. www.geographynetwork.com

GIS Data Depot. Excellent source for basemap data. www.gisdatadepot.com

Halverson, Margo. 1999. Design sense for presentations. CD–ROM. Portland, Maine: Proximity Learning.

Iliffe, Jonathan C. 2000. Datums and map projections. crcpress.com. CRC Press LLC.

Kinsley, Michael J. 1997. *Economic renewal guide: A collaborative process for sustainable community development.* Snowmass, Colo.: Rocky Mountain Institute.

Mitchell, Andy. 1999. *The ESRI guide to GIS analysis: Geographic patterns and relationships.* Vol. 1. Redlands, Calif.: ESRI Press.

Monmonier, Mark. 1996. *How to lie with maps.* Chicago and London: University of Chicago Press.

National Atlas. Download a wide variety of data for the United States (e.g., aquifers). www.nationalatlas.gov

Terraserver. An excellent source for digital orthophotographs. www.terraserver.microsoft.com

The Orton Family Foundation Community Mapping Program. 2003. *Making community connections.* Redlands, Calif.: ESRI Press.

Tufte, Edward. 1990. *Envisioning information.* Cheshire, Conn.: Graphics Press.

U.S. Department of the Interior. U.S. Geological Survey. A comprehensive source for national data sets (e.g., hydrography, elevation, land cover), digital orthophotographs, paper topographic maps, and digital elevation models. www.usgs.gov

MODULE 1: GIS BASICS

Davis, David E. 2003. *GIS for everyone.* 3d ed. Redlands, Calif.: ESRI Press.

GIS.com. An Internet guide to geographic information systems. www.gis.com

Kennedy, Heather, ed. 2001. *The ESRI Press dictionary of GIS terminology.* Redlands, Calif.: ESRI Press.

Longley, Paul, et al. 2001. *Geographic information systems and science.* Chichester, England: John Wiley & Sons.

Madej, Ed. 2001. *Cartographic design using ArcView GIS.* Clifton Park, N.Y.: OnWord Press.

Mitchell, Andy. 1998. *Zeroing in: Geographic information systems at work in the community.* Redlands, Calif.: ESRI Press.

Ormsby, Tim, et al. 1998. *Getting to know ArcView GIS.* 3d ed. Redlands, Calif.: ESRI Press.

Plewe, Brandon. 1997. *GIS online: Information retrieval, mapping, and the Internet.* Santa Fe, N.M.: OnWord Press.

Tyner, Judith. 1992. *Introduction to thematic cartography.* Upper Saddle River, N.J.: Prentice Hall, Inc.

U.S. Department of the Interior. U.S. Geological Survey. National Mapping Information. Many valuable resources and links. mapping.usgs.gov

Wood, Denis. 1992. *The power of maps.* New York: Guilford Press.

MODULE 2: REDUCING CRIME

Amdahl, Gary. 2001. *Disaster response: GIS for public safety.* Redlands, Calif.: ESRI Press.

ESRI. 2002. GIS solutions for homeland security. CD–ROM.

ESRI. 2002. GIS solutions for law enforcement. 2d ed. CD–ROM.

ESRI. Introduction to ArcView Spatial Analyst. ESRI Virtual Campus course. campus.esri.com

Green, R. W. 2000. "Health and safety." Chap. 2 in *GIS in public policy.* Redlands, Calif.: ESRI Press.

Hirschfield, Alex, and Kate Bowers. 2001. *Mapping and analysing crime data.* Taylor & Francis.

National Institute of Justice. Mapping and Analysis for Public Safety. Comprehensive Web site on issues related to crime mapping. www.ojp.usdoj.gov/nij/maps

U.S. Department of Justice. Office of Justice Programs. 1999. National Institute of Justice Research Report. Mapping crime: Principle and practice. Keith Harries. NCJ178919. Washington, D.C.

MODULE 3: A WAR ON WEEDS

Bossard, Carla C., et al. 2000. *Invasive plants of California's wildlands.* Berkeley, Calif.: University of California Press.

Federal Interagency Committee for the Management of Noxious and Exotic Weeds. 1998. *Invasive plants: Changing the landscape of America: Fact book.* Randy G. Westbrooks. Washington, D.C.

Northern Prairie Wildlife Research Center. Home page.
www.npwrc.usgs.gov/resource/2000/homing/homing.htm

Randall, John M., and Janet Marinelli. 1997. *Invasive plants: Weeds of the global garden.* 21st Century Gardening Series. Handbook No. 149. New York: Brooklyn Botanic Garden.

Staples, George W., and Robert H. Cowie. 2001. *Hawaii's invasive species: A guide to invasive plants and animals in the Hawaiian Islands (Hawaii Biological Survey handbook).* Honolulu: Bishop Museum.

U.S. Department of Agriculture. Animal and Plant Health Inspection Service. 1989. Biological control of leafy spurge.

U.S. Department of the Interior. Bureau of Land Management. Plant Conservation Alliance. Contains information about funding sources and e-mail lists on plant-related projects. www.nps.gov/plants

U.S. Department of the Interior. National Park Service. The National Park Service Wildlife and Plants. Links to various pages on invasive plants. www1.nature.nps.gov/wv

U.S. Department of the Interior, U.S. Department of Agriculture, Riley Memorial Foundation. 2000. Invasive species stakeholders, collecting, sharing and using information: Proceedings of a roundtable. Silver Spring, Md.

Whitson, Tom, ed. 2000. *Weeds of the West.* 9th ed. Western Society of Weed Science. University of Wyoming. ASIN: 0941570134. November 26.

MODULE 4: TRACKING WATER QUALITY

Lang, Laura. *Managing natural resources with GIS.* 1998. Redlands, Calif.: ESRI Press.

U.S. Department of the Interior. National Park Service. Water Resource Division. Information about the National Park Service's programs in water resources.
www1.nature.nps.gov/wrd

U.S. Department of the Interior. U.S. Geological Survey. 1992. Are fertilizers and pesticides in the ground water? A case study of the Delmarva Peninsula. Pixie A. Hamilton and Robert J. Shedlock. USGS circular 1080. Washington, D.C.: U.S. Government Printing Office.

U.S. Department of the Interior. U.S. Geological Survey. 1995. Monitoring the water quality of the nation's large rivers, Columbia River Basin NASQAN Program. USGS fact sheet 004-98. Washington, D.C.: U.S. Government Printing Office.

U.S. Department of the Interior. U.S. Geological Survey. 1999. The quality of our nation's waters: Nutrients and pesticides. USGS circular 1225. Washington, D.C.: U.S. Government Printing Office.

U.S. Department of the Interior. U.S. Geological Survey. 2001. Ground-water-level monitoring and the importance of long-term water-level data. Charles J. Taylor and W. M. Alley. USGS circular 1217. Washington, D.C.: U.S. Government Printing Office.

U.S. Department of the Interior. U.S. Geological Survey. March 2001. A primer on water quality. USGS fact sheet 027-01. Washington, D.C.: U.S. Government Printing Office.

U.S. Environmental Protection Agency. Information about EPA water topics.
www.epa.gov/ebtpages/water.html

MODULE 5: INVESTIGATING POINT-SOURCE POLLUTION

ACTION Center. The basics of landfills. Landfill information and resources. www.ejnet.org/landfills

HowStuffWorks. How landfills work. Comprehensive and informative Web site. www.howstuffworks.com/landfill.htm

Scorecard. This Environmental Defense Fund site allows you to enter your ZIP Code and get lists of pollutants currently being released into your community, quantities of those pollutants, and the entities responsible. www.scorecard.org

Tammemagi, H. Y. 1999. *The waste crisis: Landfills, incinerators, and the search for a sustainable future.* Oxford University Press.

U.S. Environmental Protection Agency. Information on environmental issues, legislation, and downloadable data. www.epa.gov

Zero Waste America. Landfills: Hazardous to the environment. Resources and links on many aspects of landfills. www.zerowasteamerica.org

MODULE 6: GETTING KIDS TO SCHOOL

ESRI. Introduction to ArcView Network Analyst. ESRI Virtual Campus course. campus.esri.com

Graham, Derek. 1993. A GIS for bus routing saves money, worry in North Carolina. *Geo Info Systems* 3 (issue 5, May).

Lang, Laura. 1999. *Transportation GIS.* Redlands, Calif.: ESRI Press.

Miller, Harvey, and Shih-Lung Shaw. 2001. *Geographic information systems for transportation: Principles and applications.* Oxford University Press.

STN Media Co., Inc. School Transportation News Online. Copyright 1998–2002. www.stnonline.com/stn/index.shtml

MODULE 7: PROTECTING THE COMMUNITY FOREST

American Forests. Many urban forestry resources and links, CITYgreen demo, and information. www.americanforests.org

Forestry Suppliers, Inc. Good source of tree inventory supplies. www.forestry-suppliers.com

Green Map System. Green maps for healthy cities and sustainable communities. www.greenmap.com

Little, Elbert J. Jr., et al. *National Audubon Society field guide to North American trees: Eastern region.* 1980. New York: Alfred A. Knopf.

National Arbor Day Foundation, The. Extensive resources and links. Copyright 2002 The National Arbor Day Foundation. www.arborday.org

Petrides, George A. 1958. *A field guide to trees and shrubs.* Peterson Field Guides. Boston, New York: Houghton Mifflin Company.

Project for Public Spaces, Inc. Urban Parks Online Suggested Reading list. Copyright 2001 Project for Public Spaces, Inc. pps.org//upo/biblio

State of Rhode Island. Rhode Island Urban and Community Forestry Plan. May 1999. Report Number 97. State Guide Plan Element 156, page 3.2.

MODULE 8: SELECTING THE RIGHT LOCATION

Colorado Division of Wildlife. Information about wildlife issues in Colorado, including current headlines and announcements. wildlife.state.co.us

ESRI. Conservation GIS using ArcView 3.x. ESRI Virtual Campus course. campus.esri.com

ESRI. Introduction to Urban and Regional Planning using ArcView 3.x. ESRI Virtual Campus course. campus.esri.com

Kinsley, Michael J. 1997. *Economic renewal guide: A collaborative process for sustainable community development.* Snowmass, Colo.: Rocky Mountain Institute.

Lagro, James A. 2001. *Site analysis: Linking program and concept in land planning and design.* New York: John Wiley & Sons.

The Orton Family Foundation. A nonprofit organization dedicated to working with youth and community partners to use GIS technology to solve community problems. Web site contains an extensive gallery of completed projects. www.orton.org

U.S. Department of Homeland Security. Federal Emergency Management Agency. Local flood-zone maps available for download. www.msc.fema.gov

Windows data installation guide

INSTALLING THE DATA ON A WINDOWS PLATFORM

Community Geography: GIS in Action includes one CD–ROM with ArcView 3.x projects and data for the exercises. To use the projects and data, ArcView 3.x must already be installed on your computer. The projects and data can be used with ArcView 3.0a through ArcView 3.3 software. The data has been tested on the following Microsoft Windows operating systems: Windows 95, Windows 98, Windows NT, Windows 2000, and Windows XP. The data takes up about 30 MB of hard-disk space. When you install the data, the entire process will take about two minutes.

Follow the steps below to install the exercise data. Do not copy the files directly from the CD to your hard drive.

1 Put the CD in your computer's CD–ROM drive. After a few moments, the CD auto-runs and the Welcome message appears. Read it. Note: If the CD does not start automatically, browse to it and run Setup.exe.

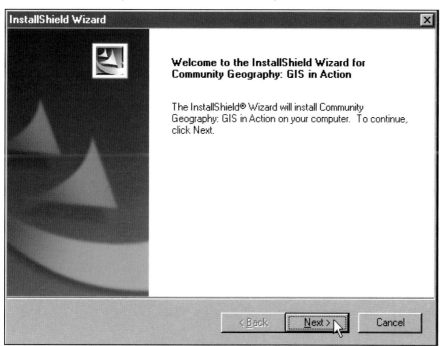

2 Click Next.

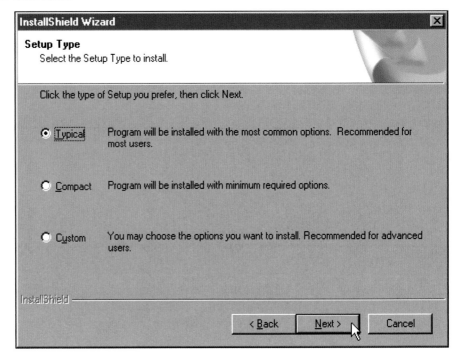

3 Click Next to complete the Typical installation.

If you choose Custom installation to change the location of where the data is installed, be sure to note this different location. Exercises in the book reference the default location *C:\esri\comgeo.*

4 Setup automatically installs the data on your computer in a folder called *comgeo.* When the data installation is complete, click Finish. Remove the CD.

ESRI provides no technical support for this complimentary CD. If you have questions or encounter problems during the installation process, refer to this complete installation guide or go to the *Community Geography: GIS in Action* Web site at *www.esri.com/communitygeography.*

UNINSTALLING THE DATA AND SOFTWARE

To uninstall the exercise data from your computer, open your operating system's control panel and click the Add/Remove Programs icon. In the Add/Remove Programs Properties dialog, select the following entry to uninstall and follow the prompts to remove it: *Community Geography: GIS in Action.*

Macintosh data installation guide

INSTALLING THE DATA ON THE MACINTOSH PLATFORM

Community Geography: GIS in Action includes one CD–ROM with data and ArcView 3.x projects for the exercises. You must have ArcView 3.0a already installed on your computer in order to use the projects and data. This guide will help you install the ArcView projects and exercise data on an Apple Macintosh computer. The data takes up approximately 30 MB of hard-disk space.

Follow the steps below to install the exercise data. Do not copy the files directly from the CD to your hard drive.

1 Put the CD in your computer's CD drive. After a few moments, you will see the *Community Geography: GIS in Action* title. Click Continue. Note: if the CD does not start automatically, browse to it and run MacInstall.

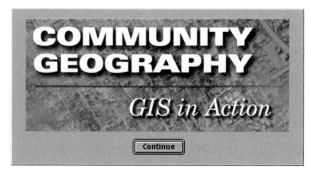

2 The Welcome message appears. Read the installation guidelines. When finished, click Install.

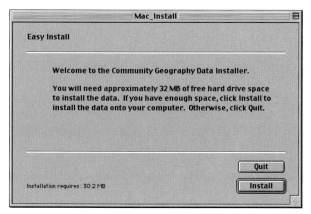

3 Navigate to the ESRI folder on your hard drive and select it. Click Choose to install the data.

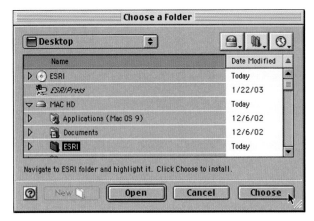

Notes: If you choose to install the data in a different folder, be sure to note the new location. Exercises in the book reference the Windows default location of *C:\esri\comgeo.* If you do not have an ESRI folder on your hard drive, you need to create one first.

If your computer hard drive has multiple partitions (e.g., computers running Mac OS X and OS 9 on separate partitions), be sure to install the Community Geography Data to the same partition where ArcView is installed.

4 A status bar displays the installation's progress. A message appears when installation is complete. Click Quit to exit the installation.

UNINSTALLING THE DATA

To uninstall the exercise data from your computer, drag the *ComGeo* folder to the trash.

Macintosh technical guide

In this book, all the illustrations show how ArcView software looks under
the Microsoft Windows operating systems. The appearance and operation of
ArcView is nearly identical on both Windows and Apple Macintosh platforms.
However, there are some cosmetic differences and a few procedural differences.
This document describes the main differences that Macintosh users will encounter. When a Macintosh user's screen differs from the screen illustrated in this
book for a particular step, reference to the information below will clarify and
reconcile the difference.

1 *No application window.* The Windows software uses an application window
   that contains all the smaller windows. On Macintosh, each window (the
   Project window and the View window as seen below) seems independent on
   the desktop.

Macintosh

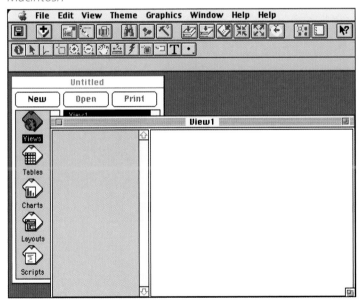

Windows

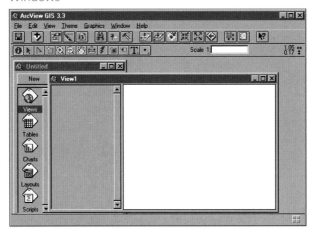

2 *Location of status bar.* ArcView software provides a status bar that explains the function of the tools and buttons. In the Macintosh environment, the status bar is located under the toolbar, near the top of the screen. The illustrations in the book show the status bar at the bottom of the ArcView application window.

Macintosh

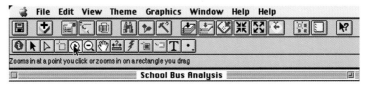

Windows

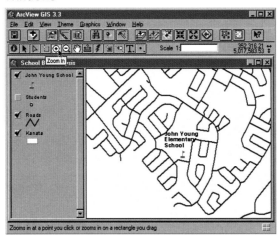

3 *Working in the active window.* If you want to perform an action in a window, you first must make it the active or frontmost window and then click again in the window to perform the action. Thus, it may seem that you need to double-click an item, even though the instructions indicate a single click. It is helpful to know which is the active window at each step of the exercise. On a Macintosh, the active window will have lines or stripes across the title bar and the nonactive windows will be blank (white).

Active window *Nonactive window*

On Windows, a single click in a nonactive window will bring that window to the front and perform the desired action, while on Macintosh a single click will only bring that window to the front. For example, in the Windows environment, if a table is behind a view and you click on the table, two things happen: the table comes to the front and the record that you click is selected. In the Macintosh environment, only one thing happens: the table comes to the front.

4 *Adding themes.* When adding themes, the window that appears is cosmetically different between Macintosh and Windows. On a Macintosh, you will find the folders/directory on the left side and the data sources on the right side of the Add Theme window. Illustrations in the book will show the list of data sources on the left side and the folders/directory structure on the right side of the Add Theme window. Also, in Macintosh, you click Add or Add All to add data. In Windows, you click OK to add the data.

Macintosh

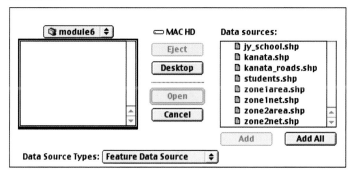

Windows

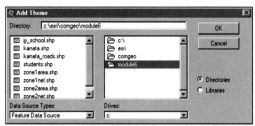

5 *Adding themes.* In the Add Theme window, Macintosh users should use Apple-click to select multiple, nonadjacent items without choosing the items in between. In the Windows version, Shift-click selects multiple, nonadjacent items without affecting intermediate items.

6 *Maximizing and closing windows.* On Macintosh, you close a window by clicking the button in the upper left corner of the window. To maximize or unmaximize a window, click the button in the upper right corner. On Windows, you can minimize, maximize, or close a window by clicking one of the three buttons located in the upper right corner of each window.

| *Macintosh* | | *Windows* | | |
| Close | Maximize | Minimize | Maximize | Close |

7 *Resizing windows.* On Macintosh, you can press and drag the bottom right corner of most windows to perform custom resizing of windows. On Windows, you can press and drag any edges or corners of most windows to perform custom resizing.

8 *Help menu.* On Macintosh, two "Help" menus are visible. The left menu is for the ArcView software and the right menu is for the Macintosh operating system. On Windows, one "Help" menu is visible, which is for the ArcView software (see the graphic under item 4 above).

9 *Pathways and folders.* When looking at a theme's properties, Macintosh indicates a change in folder using a colon (:) character, while Windows uses a backslash (\) character.

Macintosh navigation pathway
Macintosh HD:esri:comgeo:module1:map.apr

Windows navigation pathway
C:\esri\comgeo\module1\module1.apr

10 *Graphic export file formats.* When exporting a map to a graphic file, Macintosh users can export to a variety of formats, including the PICT format (compatible with most other Macintosh applications), while Windows users can export to BMP, WMF, and JPEG formats.

Data license agreement

Important: Read carefully before opening the sealed media package

Environmental Systems Research Institute, Inc. (ESRI), is willing to license the enclosed data and related materials to you only upon the condition that you accept all of the terms and conditions contained in this license agreement. Please read the terms and conditions carefully before opening the sealed media package. By opening the sealed media package, you are indicating your acceptance of the ESRI License Agreement. If you do not agree to the terms and conditions as stated, then ESRI is unwilling to license the data and related materials to you. In such event, you should return the media package with the seal unbroken and all other components to ESRI.

ESRI License Agreement

This is a license agreement, and not an agreement for sale, between you (Licensee) and Environmental Systems Research Institute, Inc. (ESRI). This ESRI License Agreement (Agreement) gives Licensee certain limited rights to use the data and related materials (Data and Related Materials). All rights not specifically granted in this Agreement are reserved to ESRI and its Licensors.

Reservation of Ownership and Grant of License: ESRI and its Licensors retain exclusive rights, title, and ownership to the copy of the Data and Related Materials licensed under this Agreement and, hereby, grant to Licensee a personal, nonexclusive, nontransferable, royalty-free, world-wide license to use the Data and Related Materials based on the terms and conditions of this Agreement. Licensee agrees to use reasonable effort to protect the Data and Related Materials from unauthorized use, reproduction, distribution, or publication.

Proprietary Rights and Copyright: Licensee acknowledges that the Data and Related Materials are proprietary and confidential property of ESRI and its Licensors and are protected by United States copyright laws and applicable international copyright treaties and/or conventions.

Permitted Uses:
Licensee may install the Data and Related Materials onto permanent storage device(s) for Licensee's own internal use.

Licensee may make only one (1) copy of the original Data and Related Materials for archival purposes during the term of this Agreement unless the right to make additional copies is granted to Licensee in writing by ESRI.

Licensee may internally use the Data and Related Materials provided by ESRI for the stated purpose of GIS training and education.

Uses Not Permitted:
Licensee shall not sell, rent, lease, sublicense, lend, assign, time-share, or transfer, in whole or in part, or provide unlicensed Third Parties access to the Data and Related Materials or portions of the Data and Related Materials, any updates, or Licensee's rights under this Agreement.

Licensee shall not remove or obscure any copyright or trademark notices of ESRI or its Licensors.

Term and Termination: The license granted to Licensee by this Agreement shall commence upon the acceptance of this Agreement and shall continue until such time that Licensee elects in writing to discontinue use of the Data or Related Materials and terminates this Agreement. The Agreement shall automatically terminate without notice if Licensee fails to comply with any provision of this Agreement. Licensee shall then return to ESRI the Data and Related Materials. The parties hereby agree that all provisions that operate to protect the rights of ESRI and its Licensors shall remain in force should breach occur.

Disclaimer of Warranty: The Data and Related Materials contained herein are provided "as-is," without warranty of any kind, either express or implied, including, but not limited to, the implied warranties of merchantability, fitness for a particular purpose, or noninfringement. ESRI does not warrant that the Data and Related Materials will meet Licensee's needs or expectations, that the use of the Data and Related Materials will be uninterrupted, or that all nonconformities, defects, or errors can or will be corrected. ESRI is not inviting reliance on the Data or Related Materials for commercial planning or analysis purposes, and Licensee should always check actual data.

Data Disclaimer: The Data used herein has been derived from actual spatial or tabular information. In some cases, ESRI has manipulated and applied certain assumptions, analyses, and opinions to the Data solely for educational training purposes. Assumptions, analyses, opinions applied, and actual outcomes may vary. Again, ESRI is not inviting reliance on this Data, and the Licensee should always verify actual Data and exercise their own professional judgment when interpreting any outcomes.

Limitation of Liability: ESRI shall not be liable for direct, indirect, special, incidental, or consequential damages related to Licensee's use of the Data and Related Materials, even if ESRI is advised of the possibility of such damage.

No Implied Waivers: No failure or delay by ESRI or its Licensors in enforcing any right or remedy under this Agreement shall be construed as a waiver of any future or other exercise of such right or remedy by ESRI or its Licensors.

Order for Precedence: Any conflict between the terms of this Agreement and any FAR, DFAR, purchase order, or other terms shall be resolved in favor of the terms expressed in this Agreement, subject to the government's minimum rights unless agreed otherwise.

Export Regulation: Licensee acknowledges that this Agreement and the performance thereof are subject to compliance with any and all applicable United States laws, regulations, or orders relating to the export of data thereto. Licensee agrees to comply with all laws, regulations, and orders of the United States in regard to any export of such technical data.

Severability: If any provision(s) of this Agreement shall be held to be invalid, illegal, or unenforceable by a court or other tribunal of competent jurisdiction, the validity, legality, and enforceability of the remaining provisions shall not in any way be affected or impaired thereby.

Governing Law: This Agreement, entered into in the County of San Bernardino, shall be construed and enforced in accordance with and be governed by the laws of the United States of America and the State of California without reference to conflict of laws principles. The parties hereby consent to the personal jurisdiction of the courts of this county and waive their rights to change venue.

Entire Agreement: The parties agree that this Agreement constitutes the sole and entire agreement of the parties as to the matter set forth herein and supersedes any previous agreements, understandings, and arrangements between the parties relating hereto.